# Picturing the Uncertain World

# Picturing the Uncertain World

How to Understand, Communicate,
and Control Uncertainty through
Graphical Display

**Howard Wainer**

PRINCETON UNIVERSITY PRESS
*Princeton & Oxford*

Published by Princeton University Press, 41 William Street,
Princeton, New Jersey 08540

In the United Kingdom: Princeton University Press, 6 Oxford Street,
Woodstock, Oxfordshire OX20 1TW



Library of Congress Cataloging-in-Publication Data

Wainer, Howard.
Picturing the uncertain world : how to understand, communicate, and control
uncertainty through graphical display / Howard Wainer.
p. cm.
Includes bibliographical references and index.
ISBN 978-0-691-13759-9 (cloth : alk. paper)
1. Uncertainty (Information theory)—Graphic methods. 2. Communication
in science—Graphic methods. I. Title.
Q375.W35 2009
003′.54—dc22 2008053529

British Library Cataloging-in-Publication Data is available

This book has been composed in Adobe Jenson Pro and DIN Pro

Printed on acid-free paper. ∞

press.princeton.edu

Printed in the United States of America

1 3 5 7 9 10 8 6 4 2

To my colleagues, past and present,
who contributed their ideas
and their kind thoughts.
This book is the result of
the beauty of their minds
and the labor of mine.

**Sine quibus non**

# Contents



In this chapter we nominate De Moivre's[1] description of the expected variation in the arithmetic mean for the title of the most dangerous equation. To support this conclusion we describe five separate examples where ignorance of this equation has led to enormous wastes of time, money, and human resources. These five examples span almost a thousand years and areas as diverse as monetary policy, education policy, medical practice, and the genetic basis of sex differences in intelligence.

## II. Political Issues

In this section we show how five different kinds of issues that emerged from essentially political arguments could be illuminated with more careful thought and a graph or two. In chapter 6, we introduce a very simple probabilistic model that yields surprising richness of understanding, which apparently escaped the editorial writers of the *New York Times*.

## V. History

We understand best those things we see grow from their very beginnings.

—Aristotle, *Metaphysics*

# Preface and Acknowledgments

The French have an expression *"J'étais marrié après l'âge de raison, mais avant l'âge de la connaissance."* This translates as "I was married after the age of reason, but before the age of knowing." *L'âge de la connaissance* is a deep idea, and hubris sometimes traps us into thinking that, while last year we hadn't quite reached l'âge de la connaissance, this year we have. This has happened to me many times in the past, but each time when a year had passed I realized that the age of knowing still lurked, just out of reach, in the future.

As I sit here, in my sixty-fifth year, I am writing this book on understanding uncertainty, because I realize that many other people share with me the fact that a search for understanding, for knowing, is a constant force in their lives. This search manifests itself in many ways, for the world we live in is filled with subtle uncertainty that can be interpreted in different ways. Among optimists, it is merely an intriguing mystery to be unraveled. Others, more cynical, see evil forces combining to trick you and thence lure you into trouble. Einstein's famous observation that "God is subtle, but not malevolent" clearly places him among the optimists. Among its deepest and most basic tenets, statistics, the science of uncertainty, follows Einstein's lead.

You don't have to be a statistician to be concerned with uncertainty; although training in statistical thinking is an invaluable aid in navigating this uncertain world. Not too long ago Sid Franks, an old college buddy and my financial planner, told me that he believes that if he did his job to perfection, the check to the undertaker would bounce. When I asked how he engineered the planning necessary to achieve this remarkably precise conclusion, he said that he plans for a life span of ninety-five years and hopes that I make it.

"Why ninety-five?" I thought. "Since the likelihood of making it to ninety-five is so small, wouldn't there be money to spare after the undertaker was paid? And wouldn't I get more enjoyment out of my money by spending it myself?"

Sid agreed, explaining that it was a tough prediction and said that he "could do a much better job" if I would just tell him well in advance exactly when it was that I was going to expire.

He understood that it is rare that such preknowledge is possible and so he had to make his plans under uncertainty. He emphasized that it is a more serious error to live beyond your money than it is to have some money left over after you die. Hence he used an upper bound for the life span.

I asked if knowing my death date was all he needed to be able to manage my account satisfactorily. He replied that, no, it would also help to know what the rate of return on my portfolio would be over the next two decades, as well as the path that inflation will take. He also pointed out that all these things change, and that the amount and direction of those changes can have a profound effect. He then emphasized the importance of having a plan as well as the flexibility to modify it as conditions change and we gain more information.

This one task contains two of the most vexing elements found in most serious decisions: uncertainty about future conditions and an asymmetric cost function (it is worse to outlive your money than to leave some over). And it is but one of thousands in all our lives that illustrate how important it is to understand and manage the uncertainty that we will inevitably encounter. But to manage and understand uncertainty, so that we can plan appropriately, we must always be gathering more data so that we test and perhaps modify our plans. This is as true of buying a house and raising a family as it is for planning our retirements. And if we do this poorly the results can be very serious indeed.

This book is about the science of uncertainty and is filled with stories that provide real life lessons on recognizing and managing uncertainty. I have accumulated these stories over twenty or so years, and in those years many debts have accumulated. It is my pleasure now to acknowledge those debts and to express my gratitude to the colleagues whose thoughts and labor are contained here.

First, my gratitude to my employer, the National Board of Medical Examiners, and its management, especially Don Melnick, Ron Nungester, and Brian Clauser, who have enthusiastically allowed me to pursue the issues described here. My special thanks to Editha Chase, who has been my right hand for the past seven years in most of my scholarly endeavors. Also, my gratitude to Caroline Derbyshire, whose sharp eyes are much appreciated.

Next to my coauthors and advisors who have generously allowed me to use our joint work here: Jacques Bertin, Henry Braun, Brian Clauser, Michael Friendly, Andrew Gelman, Marc Gessaroli, Dan Koretz, William Skorupski, Stephen Stigler, Edward Tufte, John Tukey, Danielle Vasilescu, Monica Verdi, and Harris Zwerling.

Much of the material contained here appeared initially in *Chance* magazine under the editorship serially of Hal Stern, Dalene Stangl, and Michael Lavine. I owe all three of them my thanks for the help they provided in shaping the material to be both compact and comprehensible as well as for their enthusiasm for the topics. Chapter 13 appeared in an earlier form, in the *American Statistician,* and I would like to thank the editor, Kinley Larntz, for his help.

My appreciation to the American Statistical Association for giving me permission to rework and then reuse all of this material. A complete list of the source material follows the references.

The first draft of this book owes much to the uncertain weather of the Adirondack Mountains and to the realization that activities there are limited when it rains. During one such rainy period in the summer of 2007 I sequestered myself from the rest of my family and began putting pixel to screen. During dinner each night I would regale my wife and son with the product of my activities. As a tribute to either their love or the dearth of more interesting topics, they listened intently and offered their suggestions. They added clarity, completeness, quintessence, quiet, spirituality, serenity, transcendence, detachment, intelligence, sensitivity, consonance, and grammar, while at the same time subtracting incoherence, noise, schmutz, schmerz, anxiety, catharsis, accident, grotesquerie, palpitations, exploitation, and humbuggery. I thank them.

Last, my gratitude to the staff of Princeton University Press, whose professionalism and expertise make of my work all it can be. *Primus inter pares* is my editor and friend Vickie Kearn, whose enthusiasm for this project motivated its swift completion.

# Picturing the Uncertain World

*The truth is rarely pure and never simple.*

—Oscar Wilde, 1895,
*The Importance of Being Earnest*, Act I

*Not only does God play dice with the universe,*
*he throws them where they cannot be seen.*

—Stephen Hawking

# PART I Introduction and Overview

How much is enough? While many consider this the most central existential question of our acquisitive age, it also underlies many real-world practical decisions that we must make.

Consider the question, introduced in the preface, of how much we need to save for a pleasant retirement. This sounds like a straightforward interrogative, but as anyone who has ever tried to answer it has discovered, we first need the answers to many subquestions, some of them seemingly beyond our capacity to answer. First, we need to assess how much money we now need annually to live in the way we want. A difficult question certainly, but accessible by adding together the costs of housing, food, medical expenses, transportation, recreation, and the other necessities of a comfortable retirement. But then we need to know how much these expenses are likely to change as time goes on. They can change simply through inflation, which is reasonably predictable (at least in the short to moderate term), but also through our changing needs and habits; perhaps we will be spending more for medical expenses and less for ski lift tickets. How they will change can be foreseen, but when they will do so is tougher. And, the big question, how long will we be living after we begin retirement? Because this last number might stretch thirty years or beyond, many things can change enough so that the world we envisioned in the plans we make today may not be even remotely recognizable. Unfortunately, the answer to our initial question depends strongly on the answers to all of these other questions, which are, alas, shrouded in uncertainty.

We live in a world of uncertainty. We are often faced with other decisions similar to that of saving for retirement, which require us to have an understanding of uncertainty. To aid in that understanding the last century has seen the development of a Science of Uncertainty (often called the Science of Statistics). In describing the basis of this science, Stephen Stigler, the eminent statistician/historian, told the 1987 graduating class of the University of Chicago about his fondness for the expression, "A man with one watch knows what time it is, a man with two watches is never sure." The expression could be misinterpreted as support for ignorance, which was quite the opposite of Stigler's main point. The correct interpretation is that the road to advancing knowledge runs through the recognition and measurement of uncertainty rather than through simply ignoring it. In physics, Newton gave us our first watch, and for a while we knew what time it was. But by the end of the nineteenth century relativity and quantum theory represented a second watch and we have not been sure since.

In what follows I will describe a number of compelling instances that illustrate how the science of uncertainty provides us both with answers to vexing questions and an indication of the irreducible uncertainty to which we must learn to accommodate. As part of this discussion I will frequently use what is perhaps the most powerful weapon in the arsenal of the science of uncertainty, graphic display. Graphs provide, quite literally, a map to guide us in this uncertain world. It never ceases to astonish me how much insight is provided simply by transforming quantitative information into a spatial representation. Two nineteenth-century American economists, the brothers Arthur and Henry Fahrquhar, put it best, "getting information from a table is like extracting sunlight from a cucumber." The visual display of quantitative phenomena is not only powerful but also a surprisingly recent invention. Its origins do not lie deep in the misty past, like the invention of the wheel or the control of fire, but instead is an eighteenth-century invention, due primarily to the wily Scot William Playfair (1759–1823).[1]

The science of uncertainty primarily relies on a rigorous and sometimes subtle way of thinking that may at first seem foreign and counterintuitive, but after a short while becomes so natural that we cannot imagine ever having thought in any other way. It is much like learning how to read or to swim, difficult tasks both; however, after acquiring those skills we cannot imagine that we were ever otherwise.

The first chapter is meant as a broad overview, for it is the story of the most dangerous equation. This example argues that a lack of understanding of the nature of uncertainty has yielded dire consequences and that documented cases of these unhappy circumstances can be traced back almost a thousand years. But the development of the science of uncertainty has not meant that the same kinds of errors are no longer made. On the contrary, new instances of the profound effect of general ignorance of it can be found almost daily in our newspapers. And so this chapter draws its examples from the twelfth-century monetary policies of Henry II, the educational reforms initiated by the Bill and Melinda Gates Foundation, the search for causes of kidney cancer, and the often heated discussions of the genetic bases of intelligence. These examples provide introductions to the subsequent sections of the book, which span history, politics, education, and a bit of methodology.

The chapters that form the basis of this book are presented as a series of minimysteries in which a problem is identified and a solution is sought. Our role in the discovery and disentangling of these mysteries is often akin to that of a modern Sherlock Holmes. And as Holmes we sometimes identify the mystery ourselves prior to attempting to solve it. But often the mysteries were identified by another, analogous to a modern Inspector Lestrade, whose solutions fell short either due to ignorance of modern methods for understanding uncertainty or because those methods were ignored or misapplied.

The areas of these mysteries fall into the four broad categories I just mentioned. Their substance within these categories varies widely from the origins of ideas for nineteenth-century novelists, to plausible remedies for ethnic differences in test scores, to the value of changing the way that states vote. In each case we act as a detective observing something that seems curious and tracking down clues to reveal what is likely to have happened. The tracking is done through the use of statistical thinking usually abetted by graphic displays. Sometimes we conclude with an answer, but more often the trail takes us further into the problem and points the way toward the kinds of additional data that would help us achieve deeper understanding.

As is almost always the case with any nontrivial investigation, the search for answers leads to other questions. As the wise ancients might have said, more often than not it's turtles all the way down.*

*Before modern celestial mechanics, the ancients conceived of the Earth being supported on Atlas' powerful shoulders. "But," they must have been asked, "what supports Atlas?" After some thought someone came up with the temporary solution that he stood on the back of an enormous tortoise. "But what holds the tortoise?" "Another, larger one." After that it was turtles all the way down.

1

# The Most Dangerous Equation

## 1.1. INTRODUCTION

What constitutes a dangerous equation? There are two obvious interpretations: some equations are dangerous if you know the equation and others are dangerous if you do not. In this chapter I will not explore the dangers an equation might hold because the secrets within its bounds open doors behind which lie terrible peril. Few would disagree that the obvious winner in this contest would be Einstein's iconic equation

$$E = MC^2 \tag{1.1}$$

for it provides a measure of the enormous energy hidden within ordinary matter. Its destructive capability was recognized by Leo Szilard, who then instigated the sequence of events that culminated in the construction of atomic bombs.

This is not, however, the direction I wish to pursue. Instead, I am interested in equations that unleash their danger, not when we know about them, but rather when we do not; equations that allow us to understand things clearly, but whose absence leaves us dangerously ignorant.* There are many plausible candidates that seem to fill the bill. But I feel that there is one that surpasses all others in the havoc wrought by ignorance of it over many centuries. It is the equation that provides us with the standard deviation of the sampling distribution of the mean; what might be called De Moivre's equation:

$$\sigma_{\bar{x}} = \sigma/\sqrt{n}\,. \tag{1.2}$$

For those unfamiliar with De Moivre's equation let me offer a brief aside to explain it.

**An Aside Explaining De Moivre's Equation**

The Greek letter $\sigma$, with no subscript, represents a measure of the variability of a data set (its standard deviation). So if we measure, for example, the heights of, say, 1000 students at a particular high school, we might find that the average height is 67 inches, but heights might range from perhaps as little as 55 inches to as much as 80 inches. A number that characterizes this variation is the standard deviation. Under normal circumstances we would find that about two-thirds of all children in the sample would be within one standard deviation of the average. But now suppose we randomly grouped the 1000 children into 10 groups of 100 and calculated the average within each group. The variation of these 10 averages would likely be

*(Continued)*

* One way to conceive of the concept "danger" is as a danger function

$$P(Y) = P(Y|x=1)\,P(x=1) + P(Y|x=0)\,P(x=0),$$

where

$Y$ = the event of looking like an idiot,
$x = 1$ is the event of knowing the equation in question,
$x = 0$ is the event of not knowing the equation.
$|$ is mathematical notation meaning "given that" so that the expression $P(Y|x=1)$ is read as "the probability of looking like an idiot given that you know the equation."

This equation makes explicit the two aspects of what constitutes a dangerous equation. It also indicates that danger is a product of the inherent danger and the probability of that situation occurring. Thus it may be very dangerous not to know that the mean is the sum of the observations divided by $n$, but since most people know it, it is overall not a very dangerous equation.

(*Continued*)

much smaller than $\sigma$ because it is likely that a very tall person in the group would be counterbalanced by a shorter person. De Moivre showed a relationship between the variability in the original data and the variability in the averages. He also showed how one could calculate the variability of the averages ($\sigma_{\bar{x}}$) by simply dividing the original variability by the square root of the number of individuals ($n$) that were combined to calculate the averages. And so the variability of the average of groups of 100 would be one-tenth that of the original group. Similarly, if we wanted to reduce the variability in half, we would need groups of four; to cut it to one-fifth we would need groups of 25, and so forth. That is the idea behind De Moivre's equation.

Why have I decided to choose this simple equation as the most dangerous? There are three reasons, related to

(1) the extreme length of time during which ignorance of it has caused confusion,
(2) the wide breadth of areas that have been misled, and
(3) the seriousness of the consequences that such ignorance has caused.

In the balance of this chapter I will describe five very different situations in which ignorance of De Moivre's equation has led to billions of dollars of loss over centuries, yielding untold hardship; these are but a small sampling; there are many more.

### 1.2. THE TRIAL OF THE PYX: SIX CENTURIES OF MISUNDERSTANDING

In 1150, a century after the Battle of Hastings, it was recognized that the king could not just print money and assign to it any value he chose. Instead the coinage's value must be intrinsic, based on the amount of precious materials in its makeup. And so standards were set for the weight of gold in coins—a guinea should weigh 128 grains (there are 360 grains in an ounce). It was recognized, even then, that coinage methods were too imprecise to insist that all coins be exactly equal in weight, so instead the king and the barons, who supplied the London Mint (an independent organization) with gold, insisted that coins when tested* in the aggregate [say one hundred at a time] conform to the regulated size plus or minus some allowance for variability [1/400 of the weight] which for one guinea would be 0.32 grains and so, for the aggregate, 32 grains). Obviously, they assumed that variability decreased proportionally to the number of coins and not to its square root.

This deeper understanding lay almost six hundred years in the future with De Moivre's 1730 exploration of the binomial distribution.[1,†] The costs of making errors are of two types. If the average of all the coins was too light, the barons were being cheated, for there would be extra gold left over after minting the agreed number of coins. This kind of error would easily have been detected and, if found, the director of the Mint would suffer grievous punishment. But if the variability were too great, it would mean that there would be an unacceptably large number of too heavy coins produced that could be collected, melted down, and recast with the extra gold going into the pockets of the minter. By erroneously allowing too much variability, the Mint could stay within the bounds specified and still make extra money by collecting the heavier-than-average coins and reprocessing them. The fact that this error was able to continue for almost six hundred years provides strong support for De Moivre's equation to be considered a strong candidate for the title of most dangerous equation.

### 1.3. LIFE IN THE COUNTRY: A HAVEN OR A THREAT

Figure 1.1 is a map of age-adjusted kidney cancer rates. The counties shaded are those counties that are in the lowest decile of the cancer distribution. We note that these healthy counties tend to be very rural, midwestern, southern, and western counties. It is both easy and tempting to infer that this outcome is directly due to the clean living of the rural lifestyle—no air pollution, no water pollution, access to fresh food without additives, etc.

Figure 1.2 is another map of age adjusted kidney cancer rates.

* The box the coins were kept in was called the Pyx, and so each periodic test was termed the Trial of the Pyx.

† "De Moivre (1730) knew that root *n* described the spread for the binomial, but did not comment more generally than that. Certainly by Laplace (1810) the more general version could be said to be known." (Stigler, personal communication, January 17, 2007.)

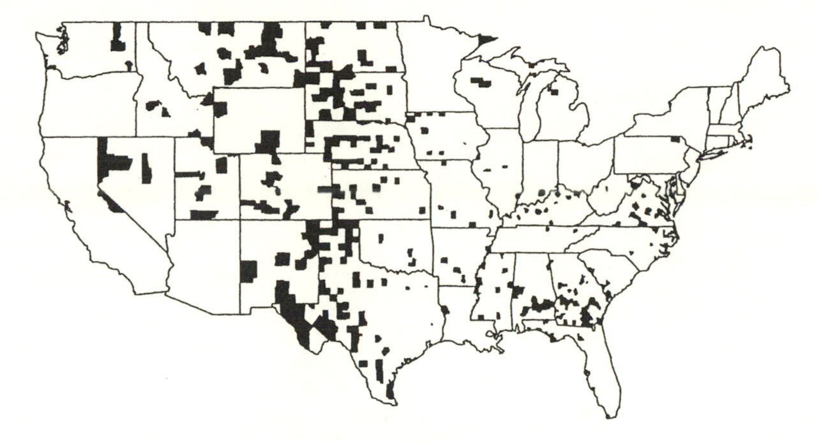

**Figure 1.1.**
Lowest kidney cancer death rates. The counties of the United States with the lowest 10% age-standardized death rates for cancer of kidney/urethra for U.S. males, 1980–1989 (from Gelman and Nolan, 2002, p. 15, reprinted with permission).

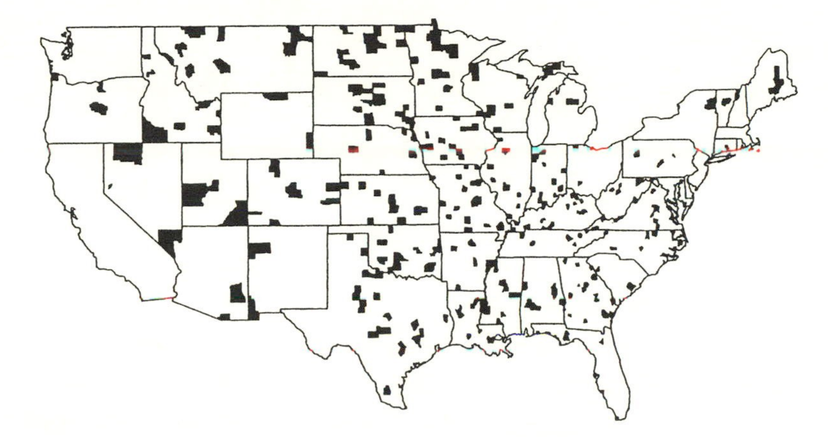

**Figure 1.2.**
Highest kidney cancer death rates. The counties of the United States with the highest 10% age-standardized death rates for cancer of kidney/urethra for U.S. males, 1980–1989 (from Gelman and Nolan, 2002, p. 14, reprinted with permission).

While it looks very much like figure 1.1, it differs in one important detail—the counties shaded are those counties that are in the *highest* decile of the cancer distribution. We note that these ailing counties tend to be very rural, midwestern, southern, and western counties. It is easy to infer that this outcome might be directly due to the poverty of the rural lifestyle—no access to good medical care, a high-fat diet, and too much alcohol, too much tobacco.

If we were to plot figure 1.1 on top of figure 1.2 we would see that many of the shaded counties on one map are right next to the shaded counties in the other. What is going on? What we are seeing is De

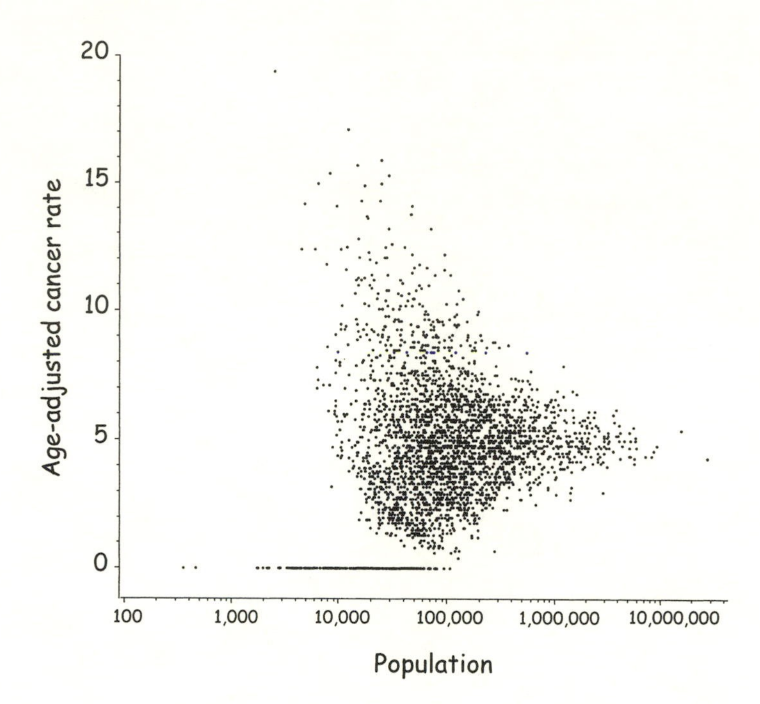


**Figure 1.3.**
Age-adjusted kidney cancer rates for all U.S. counties in 1980–1989 shown as a function of the log of the county population.

Moivre's equation in action. The variation of the mean is inversely proportional to the square root of the sample size, and so small counties have much larger variation than large counties. A county with, say one hundred inhabitants that has no cancer deaths would be in the lowest category. But if it has one cancer death it would be among the highest. Counties like New York or Los Angeles or Miami/Dade with millions of inhabitants do not bounce around like that.

If we plot the age-adjusted cancer rates against county population, this result becomes clearer still (figure 1.3). We see the typical triangular-shaped bivariate distribution in which when the population is small (left side of the graph) there is wide variation in cancer rates, from twenty per hundred thousand to zero. When county populations are large (right side of graph), there is very little variation, with all counties at about five cases per hundred thousand of population.

This is a compelling example of how someone who looked only at, say figure 1.1, and not knowing about De Moivre's equation might draw incorrect inferences and give incorrect advice (e.g., if you are at risk of kidney cancer, you should move to the wide open spaces of rural America). This would be dangerous advice, and yet this is precisely what was done in my third example.

### 1.4. THE SMALL SCHOOLS MOVEMENT: BILLIONS FOR INCREASING VARIANCE[2]

The urbanization that characterized the twentieth century yielded abandonment of the rural lifestyle and, with it, an increase in the size of schools. The time of one-room school houses ended; they were replaced by large schools, often with more than a thousand students, dozens of teachers of many specialties, and facilities that would not have been practical without the enormous increase in scale. Yet during the last quarter of the twentieth century[3] there were the beginnings of dissatisfaction with large schools, and the suggestion that smaller schools could provide better-quality education. Then in the late 1990s the Bill and Melinda Gates Foundation began supporting small schools on a broad-ranging, intensive, national basis. By 2001, the Foundation had given grants to education projects totaling approximately $1.7 billion. They have since been joined in support for smaller schools by the Annenberg Foundation, the Carnegie Corporation, the Center for Collaborative Education, the Center for School Change, Harvard's Change Leadership Group, Open Society Institute, Pew Charitable Trusts, and the U.S. Department of Education's Smaller Learning Communities Program. The availability of such large amounts of money to implement a smaller schools policy yielded a concomitant increase in the pressure to do so, with programs to splinter large schools into smaller ones being proposed and implemented broadly (e.g., New York City, Los Angeles, Chicago, and Seattle).

What is the evidence in support of such a change? There are many claims made about the advantages of smaller schools, but we will focus here on just one—that when schools are smaller, students' achievement improves. That is, the expected achievement in schools, given that they are small [$E(\text{achievement}|\text{small})$] is greater than what is expected if they are big [$E(\text{achievement}|\text{big})$]. Using this convenient mathematical notation,

$$E(\text{achievement}|\text{small}) > E(\text{achievement}|\text{big}), \tag{1.3}$$

all else being equal. But the supporting evidence for this is that, when one looks at high-performing schools, one is apt to see an unrepresentatively large proportion of smaller schools. Or, stated mathematically, that

$$P(\text{small}|\text{high achievement}) > P(\text{large}|\text{high achievement}). \quad (1.4)$$

Note that expression (1.4) does not imply (1.3).

In an effort to see the relationship between small schools and achievement we looked at the performance of all of Pennsylvania's public schools, as a function of their size, on the Pennsylvania testing program (PSSA), which is very broad and yields scores in a variety of subjects and over the entire range of precollegiate school years. If we examine the mean scores of the 1662 separate schools that provide fifth grade reading scores, we find that of the top-scoring fifty schools (the top 3%), six of them were among the smallest 3% of the schools. This is an overrepresentation by a factor of four. If size of school was unrelated to achievement, we would expect 3% of small schools to be in this select group, and we found 12%. The bivariate distribution of enrollment and test score is shown in figure 1.4. The top fifty schools are marked by a square.

We also identified the fifty lowest scoring schools, marked by an "o" in figure 1.4. Nine of these (18%) were among the fifty smallest schools. This result is completely consonant with what is expected from De Moivre's equation—the smaller schools are expected to have higher variance and hence should be overrepresented at both extremes. Note that the regression line shown on figure 1.4 is essentially flat, indicating that, overall, there is no apparent relationship between school size and performance. But this is not always true.

Figure 1.5 is a plot of eleventh grade scores. We find a similar overrepresentation of small schools on both extremes, but this time the regression line shows a significant positive slope; overall, students at bigger schools do better.

The small schools movement seems to have arrived at one of its recommendations through the examination of only one tail of the performance distribution. Small schools are overrepresented at both tails, exactly as expected, since smaller schools will show greater variation in performance and empirically will show up wherever we look. Our

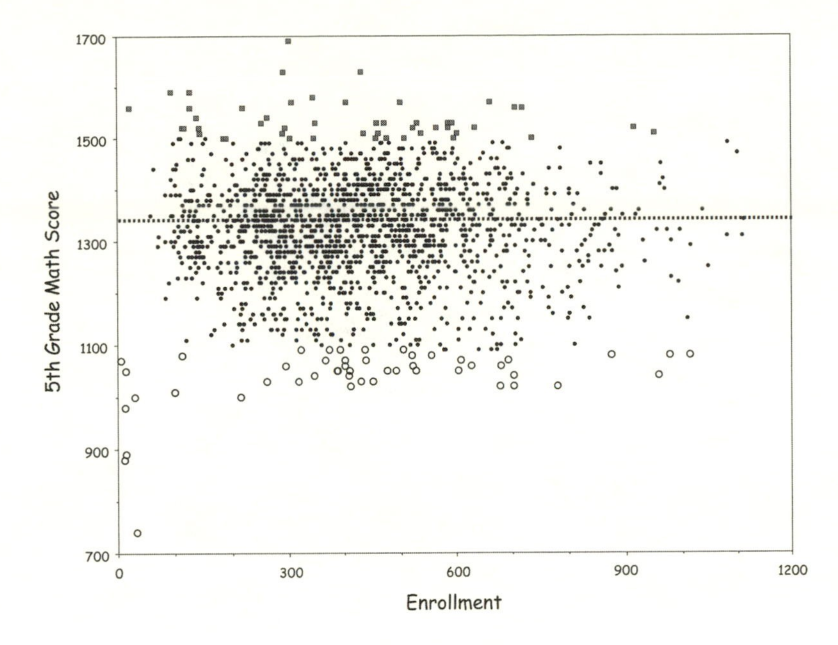


**Figure 1.4.** Average score of fifth grade classes in mathematics shown as a function of school size.

examination of fifth grade performance suggests that school size alone seems to have no bearing on student achievement. This is not true at the high-school level, where larger schools show better performance. This too is not unexpected, since very small high schools cannot provide as broad a curriculum or as many highly specialized teachers as large schools. This was discussed anecdotally in a July 20, 2005, article in the *Seattle Weekly* by Bob Geballe. The article describes the conversion of Mountlake Terrace High School in Seattle from a large suburban school with an enrollment of 1,800 students into five smaller schools. The conversion was greased with a Gates Foundation grant of almost a million dollars. Although class sizes remained the same, each of the five schools had fewer teachers. Students complained, "There's just one English teacher and one math teacher . . . teachers ended up teaching things they don't really know." Perhaps this helps to explain the regression line in figure 1.5.

On October 26, 2005, Lynn Thompson, in an article in the *Seattle Times* reported that "The Gates Foundation announced last week it is moving away from its emphasis on converting large high schools into smaller ones and instead giving grants to specially selected school

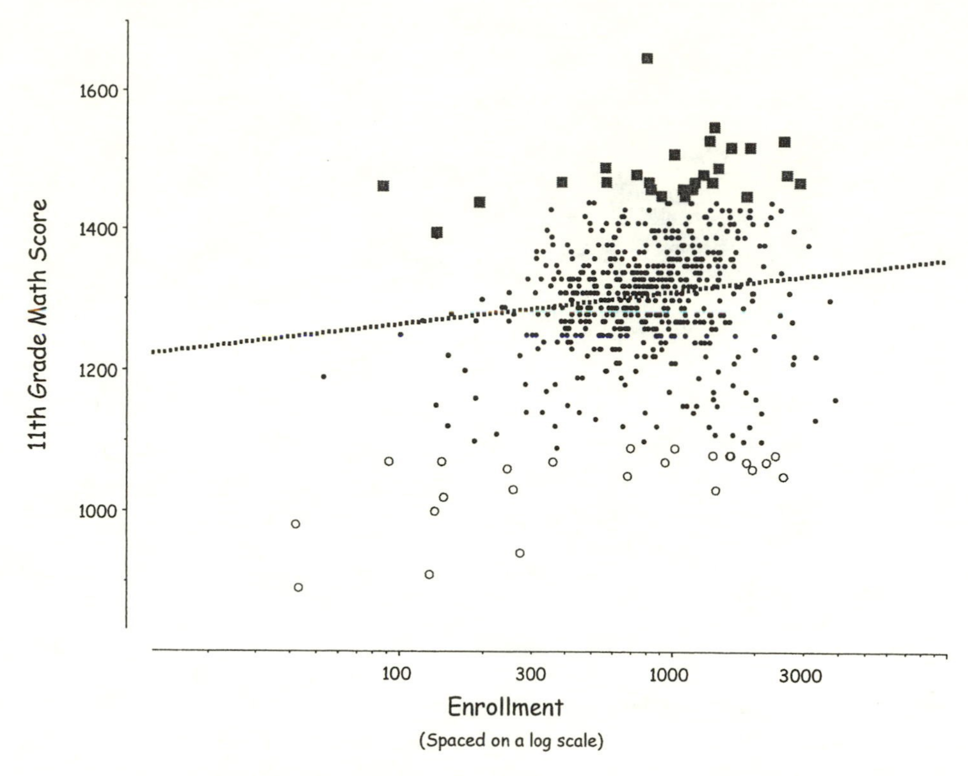


**Figure 1.5.** Eleventh grade student scores on a statewide math test shown as a function of school size.

districts with a track record of academic improvement and effective leadership. Education leaders at the Foundation said they concluded that improving classroom instruction and mobilizing the resources of an entire district were more important first steps to improving high schools than breaking down the size." This point of view was amplified in a study that carefully analyzed matched students in schools of varying sizes. The lead author concluded, "I'm afraid we have done a terrible disservice to kids."[4]

Expending more than a billion dollars on a theory based on ignorance of De Moivre's equation suggests just how dangerous that ignorance can be.

### 1.5. THE SAFEST CITIES

The *New York Times* recently reported[5] the ten safest U.S. cities and the ten most unsafe based on an automobile insurance company statistic, the "average number of years between accidents." The cities were drawn from the two hundred largest cities in the U.S. It should come as no surprise that a list of the ten safest cities, the ten most dangerous cities, and the ten largest cities have no overlap (see table 1.1).

Exactly which cities are the safest, and which the most dangerous, must surely depend on many things. But it would be difficult (because of De Moivre's equation) for the largest cities to be at the extremes. Thus we should not be surprised that the ends of the safety distribution are anchored by smaller cities.

### 1.6. SEX DIFFERENCES

For many years it has been well established that there is an overabundance of boys at the high end of test score distributions. This has meant that about twice as many boys as girls received Merit Scholarships and other highly competitive awards. Historically, some observers have used such results to make inferences about differences in intelligence between the sexes. Over the last few decades, however, most enlightened investigators have seen that it is not necessarily a difference in level but a difference in variance that separates the sexes. The public observation of this fact has not been greeted gently; witness the recent outcry when Harvard's (now ex-)president Lawrence Summers pointed this out:

> It does appear that on many, many, different human attributes—height, weight, propensity for criminality, overall IQ, mathematical ability, scientific ability—there is relatively clear evidence that whatever the difference in means—which can be debated—there is a difference in standard deviation/variability of a male and female population. And it is true with respect to attributes that are and are not plausibly, culturally determined.* (Summers [2005])

The boys' score distributions are almost always characterized by greater variance than the girls'. Thus, while there are more boys at the high end, there are also more at the low end.

*Informal remarks made by Lawrence Summers in July 2005 at the National Bureau of Economic Research Conference on Diversifying the Science and Engineering Workforce.

Table 1.1
**Information on Automobile Accident Rates in 20 Cities**

| *City* | *State* | *Population rank* | *Population* | *Number of years between accidents* |
|---|---|---|---|---|
| **Ten safest** | | | | |
| Sioux Falls | South Dakota | 170 | 133,834 | 14.3 |
| Fort Collins | Colorado | 182 | 125,740 | 13.2 |
| Cedar Rapids | Iowa | 190 | 122,542 | 13.2 |
| Huntsville | Alabama | 129 | 164,237 | 12.8 |
| Chattanooga | Tennessee | 138 | 154,887 | 12.7 |
| Knoxville | Tennessee | 124 | 173,278 | 12.6 |
| Des Moines | Iowa | 103 | 196,093 | 12.6 |
| Milwaukee | Wisconsin | 19 | 586,941 | 12.5 |
| Colorado Springs | Colorado | 48 | 370,448 | 12.3 |
| Warren | Michigan | 169 | 136,016 | 12.3 |
| **Ten least safe** | | | | |
| Newark | New Jersey | 64 | 277,911 | 5.0 |
| Washington | DC | 25 | 563,384 | 5.1 |
| Elizabeth | New Jersey | 189 | 123,215 | 5.4 |
| Alexandria | Virginia | 174 | 128,923 | 5.7 |
| Arlington | Virginia | 114 | 187,873 | 6.0 |
| Glendale | California | 92 | 200,499 | 6.1 |
| Jersey City | New Jersey | 74 | 239,097 | 6.2 |
| Paterson | New Jersey | 148 | 150,782 | 6.5 |
| San Francisco | California | 14 | 751,682 | 6.5 |
| Baltimore | Maryland | 18 | 628,670 | 6.5 |
| **Ten biggest** | | | | |
| New York | New York | 1 | 8,085,742 | 8.4 |
| Los Angeles | California | 2 | 3,819,951 | 7.0 |
| Chicago | Illinois | 3 | 2,869,121 | 7.5 |
| Houston | Texas | 4 | 2,009,690 | 8.0 |
| Philadelphia | Pennsylvania | 5 | 1,479,339 | 6.6 |
| Phoenix | Arizona | 6 | 1,388,416 | 9.7 |
| San Diego | California | 7 | 1,266,753 | 8.9 |
| San Antonio | Texas | 8 | 1,214,725 | 8.0 |
| Dallas | Texas | 9 | 1,208,318 | 7.3 |
| Detroit | Michigan | 10 | 911,402 | 10.4 |

Source: Data from the *NY Times* and www.allstate.com/media/newsheadlines. From Wainer, 2007d.

TABLE 1.2

**Eighth Grade NAEP National Results: Summary of Some Outcomes, by Sex, from National Assessment of Educational Progress**

| | | *Mean scale scores* | | *Standard deviations* | | |
|---|---|---|---|---|---|---|
| *Subject* | *Year* | *Male* | *Female* | *Male* | *Female* | *Male/female ratio* |
| Math | 1990 | 263 | 262 | 37 | 35 | 1.06 |
| | 1992 | 268 | 269 | 37 | 36 | 1.03 |
| | 1996 | 271 | 269 | 38 | 37 | 1.03 |
| | 2000 | 274 | 272 | 39 | 37 | 1.05 |
| | 2003 | 278 | 277 | 37 | 35 | 1.06 |
| | 2005 | 280 | 278 | 37 | 35 | 1.06 |
| Science | 1996 | 150 | 148 | 36 | 33 | 1.09 |
| | 2000 | 153 | 146 | 37 | 35 | 1.06 |
| | 2005 | 150 | 147 | 36 | 34 | 1.06 |
| Reading | 1992 | 254 | 267 | 36 | 35 | 1.03 |
| | 1994 | 252 | 267 | 37 | 35 | 1.06 |
| | 1998 | 256 | 270 | 36 | 33 | 1.09 |
| | 2002 | 260 | 269 | 34 | 33 | 1.03 |
| | 2003 | 258 | 269 | 36 | 34 | 1.06 |
| | 2005 | 257 | 267 | 35 | 34 | 1.03 |
| Geography | 1994 | 262 | 258 | 35 | 34 | 1.03 |
| | 2001 | 264 | 260 | 34 | 32 | 1.06 |
| US History | 1994 | 259 | 259 | 33 | 31 | 1.06 |
| | 2001 | 264 | 261 | 33 | 31 | 1.06 |

SOURCE: http://nces.ed.gov/nationsreportcard/nde/

An example, chosen from the National Assessment of Educational Progress (NAEP), is shown in table 1.2. NAEP is a true survey and so problems of self-selection (rife in college entrance exams, licensing exams, etc.) are substantially reduced. The data summarized in table 1.2 were accumulated over fifteen years and five subjects. In all instances the standard deviation of males is from 3% to 9% greater than that of females. This is true when males score higher on average (math, science, geography) or lower (reading).

Both inferences, the incorrect one about differences in level and the correct one about differences in variability, cry out for explanation.

The old cry would have been "why do boys score higher than girls?"; the newer one, "why do boys show more variability?" If one did not know about De Moivre's result and tried to answer only the first question, it would be a wild goose chase, trying to find an explanation for a phenomenon that did not exist. But, if we focus on why variability is greater in males, we may find pay dirt. Obviously the answer to the causal question "why?" will have many parts. Surely socialization and differential expectations must be major components—especially in the past, before the realization grew that a society cannot compete effectively in a global economy with only half its workforce fully mobilized. But there is another component that is key—and especially related to the topic of this chapter, De Moivre's equation.

In discussing Lawrence Summers' remarks about sex differences in scientific ability, Christiane Nüsslein-Volhard, the 1995 Nobel Laureate in Physiology/Medicine,* said

> He missed the point. In mathematics and science, there is no difference in the intelligence of men and women. The difference in genes between men and women is simply the Y chromosome, which has nothing to do with intelligence. (Dreifus [July 4, 2006])

But perhaps it is Professor Nüsslein-Volhard who missed the point here. The *Y* chromosome is not the only difference between the sexes. Summers' point was that, when we look at either extreme of an ability distribution, we will see more of the group that has greater variation. Any mental trait that is conveyed on the *X* chromosome will have larger variability among males than females, for females have two *X* chromosomes versus only one for males. Thus, from De Moivre's equation, we would expect, *ceteris paribus*, about 40% more variability† among males than females. The fact that we see less than 10% greater variation in NAEP demands the existence of a deeper explanation. First, De Moivre's equation requires independence of the two *X*'s, and

*Dr. Nüsslein-Volhard shared the prize with Eric F. Wieschaus for their work that showed how the genes in a fertilized egg direct the formation of an embryo.

† Actually the square root of two (1.414...) is more, hence my approximation of 40% is a tad low; it would be more accurate to have said 41.4%, but "40%" makes the point.

with assortative mating this is not going to be true. Additionally, both *X* chromosomes are not expressed in every cell. Moreover, there must be major causes of high-level performance that are not carried on the *X* chromosome, and indeed are not genetic. But it suggests that for some skills between 10% and 25% of the increased variability is likely to have had its genesis on the *X* chromosome. This observation would be invisible to those, even those with Nobel prizes for work in genetics, who are in ignorance of De Moivre's equation.

It is well established that there is evolutionary pressure toward greater variation within species—within the constraints of genetic stability. This is evidenced by the dominance of sexual over asexual reproduction among mammals. But this leaves us with a puzzle. Why was our genetic structure built to yield greater variation among males than females? And not just among humans, but virtually all mammals. The pattern of mating suggests an answer. In most mammalian species that reproduce sexually, essentially all adult females reproduce, whereas only a small proportion of males do so (modern humans excepted). Think of the alpha-male lion surrounded by a pride of females, with lesser males wandering aimlessly and alone in the forest roaring in frustration. One way to increase the likelihood of off-spring being selected to reproduce is to have large variance among them. Thus evolutionary pressure would reward larger variation for males relative to females.

This view gained further support in studies by Arthur Arnold and Eric Vilain of UCLA that were reported by Nicholas Wade of the *New York Times* on April 10, 2007. He wrote,

> It so happens that an unusually large number of brain-related genes are situated on the *X* chromosome. The sudden emergence of the *X* and *Y* chromosomes in brain function has caught the attention of evolutionary biologists. Since men have only one *X* chromosome, natural selection can speedily promote any advantageous mutation that arises in one of the *X*'s genes. So if those picky women should be looking for smartness in prospective male partners, that might explain why so many brain-related genes ended up on the *X*.

He goes on to conclude,

> Greater male variance means that although average IQ is identical in men and women, there are fewer average men and more at both extremes. Women's care in selecting mates, combined with the fast selection made possible by men's lack of backup copies of *X*-related genes, may have driven the divergence between male and female brains.

### 1.7. CONCLUSION

It is no revelation that humans don't fully comprehend the effect that variation, and especially differential variation, has on what we observe. Daniel Kahneman's 2002 Nobel Prize was for his studies on intuitive judgment (which occupies a middle ground "between the automatic operations of perception and the deliberate operations of reasoning"[6]. Kahneman showed that humans don't intuitively "know" that smaller hospitals will have greater variability in the proportion of male to female births. But such inability is not limited to humans making judgments in psychology experiments. Small hospitals are routinely singled out for special accolades because of their exemplary performance, only to slip toward average in subsequent years. Explanations typically abound that discuss how their notoriety has overloaded their capacity. Similarly, small mutual funds are recommended, *post hoc*, by Wall Street analysts, only to have their subsequent performance disappoint investors. The list goes on and on, adding evidence and support to my nomination of De Moivre's equation as the most dangerous of them all.

This chapter has been aimed at reducing the peril that accompanies ignorance of De Moivre's equation and also at showing why an understanding of variability is critical if we are to avoid serious errors.

# PART II Political Issues

Nowhere is the science of uncertainty needed more than in the great public policy questions that are regularly debated. In this section I try to illustrate how these debates can be informed with careful thought, structured in a formal way. This is especially true in chapter 6, where we see how a simple mathematical model, based on the most rudimentary probability theory, yields deep insights about an important question of democratic representation—a question that confused even the editorial writers of the *New York Times*. But this is not the only time that we might be led astray by what appears on the pages of the *Times*. Chapter 3 tells of an op-ed piece by former secretary of state George Schultz that adds further confusion by violating some of the canons of the science of uncertainty. We then show (chapter 4) how the newly adopted Medicare Drug Plan can be understood visually even though its underlying structure seems to have befuddled many commentators. But graphical display does not always shine a light into the dark corners of a mysterious topic. Sometimes (chapter 5), a graphic when cleverly prepared can lie to us even more effectively than words, for lying words are much more easily forgotten than a deceitful picture. This same theme is what begins the section as we take some obvious lies and see if we can discover the truth behind them.

**Figure 2.0.** DILBERT: © Scott Adams/Dist. by United Features Syndicate, Inc.

2

# Curbstoning IQ and the 2000 Presidential Election

The government [is] extremely fond of amassing great quantities of statistics. These are raised to the *n*th degree, the cube roots are extracted, and the results are arranged into elaborate and impressive displays. What must be kept ever in mind, however, is that in every case, the figures are first put down by a village watchman, and he puts down anything he damn well pleases.

—Sir Josiah Charles Stamp (1880–1941)

In the February 1993 issue of the *Atlantic Monthly,* Charles C. Mann wrote of his experiences working as an enumerator for the 1980 U.S. Census.

> I answered an advertisement and attended a course. In a surprisingly short time I became an official enumerator. My job was to visit apartments that had not mailed back their census forms. As identification, I was given a black plastic briefcase with a big red, white, and blue sticker that said U.S. CENSUS. I am a gangling, six-foot-four-inch Caucasian; the government sent me to Chinatown.
>
> Strangely enough, I was a failure. Some people took one look and slammed the door. One elderly woman burst into tears at the sight of me. I was twice offered money to go away. Few residents had time to fill out a long form.

> Eventually I met an old census hand. "Why don't you just curbstone it?" he asked. "Curbstoning," I learned, was enumerator jargon for sitting on the curbstone and filling out forms with made-up information.

The term curbstoning describes a practice that is widespread enough to warrant inclusion in the glossary of *The 2000 Census: Interim Assessment,* which defines it as "The practice by which a census enumerator fabricates a questionnaire for a residence without actually visiting it."

There have been many studies of curbstoning, most of which focus on assessing its frequency of occurrence[1] and its plausible impact on final results. It has also figured prominently in lawsuits that challenge census findings (e.g., in 1992, The City of New York vs. United States Department of Commerce, City of Atlanta vs. Mosbacher, Florida House of Representatives vs. Barbara H. Franklin, Secretary of Commerce). Estimates about the extent of curbstoning vary, depending on neighborhood and question. A 1986 study by the Census estimated that 3% to 5% of all enumerators engaged in some form of curbstoning, and, after thorough investigation, 97% of those suspected were dismissed.

Obviously, the existence of curbstoning threatens the validity of the census, and strenuous efforts to eradicate it are fully justified. But how much does the inclusion of this sort of contaminant really affect results? The answer to this depends on two things, the incidence of curbstoning and the accuracy of the curbstone estimates. Obviously, if the incidence is less than 5%, that marks the upper bound on the size of the error. But what about the second component? The Princeton polymath John Tukey once remarked to me that, on average, curbstone estimates themselves were within 4% of their true values. Obviously this is an average figure and will depend on the specific question whose answer is being fabricated. Unfortunately, Tukey didn't elaborate. While I have not yet been successful in tracking down the source of Tukey's comment, it does match more modern estimates of the accuracy of average estimates.[2]

Unfortunately, curbstone estimates are not limited to the U.S. Census. In the spring of 1913, H. H. Goddard, the director of research at the Vineland Training School for Feeble-Minded Girls and Boys in New Jersey, directed two women to Ellis Island for more than two months. Their task was to pick out feeble-minded immigrants (estimate IQ) by sight. He pointed out that

> "After a person has had considerable experience in this work he almost gets a sense of what a feeble-minded person is so that he can tell one from afar off. . . . These two women could pick out the feeble-minded without the aid of the Binet test at all." (Goddard [1913], p. 106)

The practice of obtaining curbstone estimates of IQ apparently did not end with Goddard almost a century ago. Indeed, although it was not identified as such, a new exemplar arrived in my email box recently (see table 2.1).

The obvious intent of this data display is to provide another characterization of the states that supported George Bush (poorer, dumber) in the 2000 election versus those that supported Al Gore (smarter, richer). Even a quick look at the table brings us up short. The income data seem awfully low. What year was this for? And where did they get state IQ data? I know of no national program that provides a legitimate sample of IQ data. And could the mean IQ of Mississippi really be 85? It is reminiscent of Goddard's finding that more than 40% of Russia was feeble minded. Obviously these data could not be real.

The table was originally posted to the Mensa website by someone named Robert Calvert in 2002 and has since been identified as fake[3]—but it took a while. Why? Perhaps because the results conformed, at least qualitatively, to most readers' general expectations. To what extent are these curbstone estimates really plausible? Is the outcome, at least qualitatively, correct?

To try to answer this I gathered 2002 median state income data from U.S. Census and found that while the incomes are much larger (see table 2.2), the rank order of the states is very similar; they are correlated at .81 (the bivariate distribution* of state incomes has two outliers—the faked data have Utah much too low and New York much

* "Bivariate distribution" is statistical jargon for the array of points in the plot of true state income data, on the horizontal axis, and fake state income data on the vertical axis.

Table 2.1
**Data of Questionable Quality on State Incomes, IQs, and Voting in the 2000 Presidential Election**

| *Rank* | *State* | *Average IQ* | *Average income* | *2000 election* |
|---|---|---|---|---|
| 1 | Connecticut | 113 | $26,979 | Gore |
| 2 | Massachusetts | 111 | $24,059 | Gore |
| 3 | New Jersey | 111 | $26,457 | Gore |
| 4 | New York | 109 | $23,534 | Gore |
| 5 | Rhode Island | 107 | $20,299 | Gore |
| 6 | Hawaii | 106 | $21,218 | Gore |
| 7 | Maryland | 105 | $22,974 | Gore |
| 8 | New Hampshire | 105 | $22,934 | Bush |
| 9 | Illinois | 104 | $21,608 | Gore |
| 10 | Delaware | 103 | $21,451 | Gore |
| 11 | Minnesota | 102 | $20,049 | Gore |
| 12 | Vermont | 102 | $18,834 | Gore |
| 13 | Washington | 102 | $20,398 | Gore |
| 14 | California | 101 | $21,278 | Gore |
| 15 | Pennsylvania | 101 | $20,253 | Gore |
| 16 | Maine | 100 | $18,226 | Gore |
| 17 | Virginia | 100 | $20,629 | Bush |
| 18 | Wisconsin | 100 | $18,727 | Gore |
| 19 | Colorado | 99 | $20,124 | Bush |
| 20 | Iowa | 99 | $18,287 | Gore |
| 21 | Michigan | 99 | $19,508 | Gore |
| 22 | Nevada | 99 | $20,266 | Bush |
| 23 | Ohio | 99 | $18,624 | Bush |
| 24 | Oregon | 99 | $18,202 | Gore |
| 25 | Alaska | 98 | $21,603 | Bush |
| 26 | Florida | 98 | $19,397 | Bush |
| 27 | Missouri | 98 | $18,835 | Bush |
| 28 | Kansas | 96 | $19,376 | Bush |
| 29 | Nebraska | 95 | $19,084 | Bush |
| 30 | Arizona | 94 | $17,119 | Bush |
| 31 | Indiana | 94 | $18,043 | Bush |
| 32 | Tennessee | 94 | $17,341 | Bush |
| 33 | North Carolina | 93 | $17,667 | Bush |
| 34 | West Virginia | 93 | $15,065 | Bush |
| 35 | Arkansas | 92 | $15,439 | Bush |
| 36 | Georgia | 92 | $18,130 | Bush |

*(Continued)*

Table 2.1 (*Continued*)

| *Rank* | *State* | *Average IQ* | *Average income* | *2000 election* |
|---|---|---|---|---|
| 37 | Kentucky | 92 | $16,534 | Bush |
| 38 | New Mexico | 92 | $15,353 | Gore |
| 39 | North Dakota | 92 | $16,854 | Bush |
| 40 | Texas | 92 | $17,892 | Bush |
| 41 | Alabama | 90 | $16,220 | Bush |
| 42 | Louisiana | 90 | $15,712 | Bush |
| 43 | Montana | 90 | $16,062 | Bush |
| 44 | Oklahoma | 90 | $16,198 | Bush |
| 45 | South Dakota | 90 | $16,558 | Bush |
| 46 | South Carolina | 89 | $15,989 | Bush |
| 47 | Wyoming | 89 | $17,423 | Bush |
| 48 | Idaho | 87 | $16,067 | Bush |
| 49 | Utah | 87 | $15,325 | Bush |
| 50 | Mississippi | 85 | $14,088 | Bush |

Note: The income-IQ correlation was inspired by the book *IQ and the Wealth of Nations*, by Richard Lynn and Tatu Vanhanen. The income statistics are now perhaps a decade old, but were apparently the only numbers available to the original compiler when the results of the 2000 election became available (and this chart was made). Commentary is welcome; send to gcharter@student.umass.edu.

too high). But what about IQ? There really are no data that can serve as a proxy for IQ, but with the passage of No Child Left Behind all states must participate in the National Assessment of Educational Progress (NAEP). This has meant that, in 2003, for the first time, we have data on academic performance from what can be considered a rigorously constructed random sample of children from each state. Data are now available[4] on math and reading performance for fourth and eighth graders enrolled in public schools. It is probably reasonable to assume that academic performance data for students have some positive relation to intelligence.

The ordering of the states is essentially the same regardless of which of the four NAEP measures we might choose to use to represent that state. Thus I opted to summarize each state by a simple composite mean of these four measures; since each individual measure correlates at least at .95 with the mean, we are not doing any serious harm using it. Now we are almost ready to try to answer the question

Table 2.2
**Three-Year-Average Median Household Income by State: 2000–2002 (Income in 2002 Dollars)**

| State | Median income | Math-4 | NAEP Scores Rdg-4 | Math-8 | Rdg-8 | Mean | 2000 election |
|---|---|---|---|---|---|---|---|
| Massachusetts | $ 50,587 | 242 | 228 | 287 | 273 | 257 | Gore |
| Vermont | $ 41,929 | 242 | 226 | 286 | 271 | 256 | Gore |
| Minnesota | $ 54,931 | 242 | 223 | 291 | 268 | 256 | Gore |
| Connecticut | $ 53,325 | 241 | 228 | 284 | 267 | 255 | Gore |
| Iowa | $ 41,827 | 238 | 223 | 284 | 268 | 253 | Gore |
| New Jersey | $ 53,266 | 239 | 225 | 281 | 268 | 253 | Gore |
| Maine | $ 37,654 | 238 | 224 | 282 | 268 | 253 | Gore |
| Wisconsin | $ 46,351 | 237 | 221 | 284 | 266 | 252 | Gore |
| Washington | $ 44,252 | 238 | 221 | 281 | 264 | 251 | Gore |
| New York | $ 42,432 | 236 | 222 | 280 | 265 | 251 | Gore |
| Delaware | $ 50,878 | 236 | 224 | 277 | 265 | 250 | Gore |
| Oregon | $ 42,704 | 236 | 218 | 281 | 264 | 250 | Gore |
| Pennsylvania | $ 43,577 | 236 | 219 | 279 | 264 | 249 | Gore |
| Michigan | $ 45,335 | 236 | 219 | 276 | 264 | 249 | Gore |
| Illinois | $ 45,906 | 233 | 216 | 277 | 266 | 248 | Gore |
| Maryland | $ 55,912 | 233 | 219 | 278 | 262 | 248 | Gore |
| Rhode Island | $ 44,311 | 230 | 216 | 272 | 261 | 245 | Gore |
| Hawaii | $ 49,775 | 227 | 208 | 266 | 251 | 238 | Gore |
| California | $ 48,113 | 227 | 206 | 267 | 251 | 238 | Gore |
| New Mexico | $ 35,251 | 223 | 203 | 263 | 252 | 235 | Gore |
| New Hampshire | $ 53,549 | 243 | 228 | 286 | 271 | 257 | Bush |
| North Dakota | $ 36,717 | 238 | 222 | 287 | 270 | 254 | Bush |
| South Dakota | $ 38,755 | 237 | 222 | 285 | 270 | 254 | Bush |
| Montana | $ 33,900 | 236 | 223 | 286 | 270 | 254 | Bush |
| Wyoming | $ 40,499 | 241 | 222 | 284 | 267 | 253 | Bush |
| Virginia | $ 49,974 | 239 | 223 | 282 | 268 | 253 | Bush |
| Kansas | $ 42,523 | 242 | 220 | 284 | 266 | 253 | Bush |
| Colorado | $ 49,617 | 235 | 224 | 283 | 268 | 252 | Bush |
| Ohio | $ 43,332 | 238 | 222 | 282 | 267 | 252 | Bush |
| North Carolina | $ 38,432 | 242 | 221 | 281 | 262 | 252 | Bush |
| Nebraska | $ 43,566 | 236 | 221 | 282 | 266 | 251 | Bush |
| Indiana | $ 41,581 | 238 | 220 | 281 | 265 | 251 | Bush |
| Missouri | $ 43,955 | 235 | 222 | 279 | 267 | 251 | Bush |
| Utah | $ 48,537 | 235 | 219 | 281 | 264 | 250 | Bush |

*(Continued)*

Table 2.2 (*Continued*)

| *State* | *Median income* | *Math-4* | *NAEP Scores Rdg-4* | *Math-8* | *Rdg-8* | *Mean* | *2000 election* |
|---|---|---|---|---|---|---|---|
| Idaho | $ 38,613 | 235 | 218 | 280 | 264 | 249 | Bush |
| Kentucky | $ 37,893 | 229 | 219 | 274 | 266 | 247 | Bush |
| Texas | $ 40,659 | 237 | 215 | 277 | 259 | 247 | Bush |
| South Carolina | $ 38,460 | 236 | 215 | 277 | 258 | 246 | Bush |
| Florida | $ 38,533 | 234 | 218 | 271 | 257 | 245 | Bush |
| West Virginia | $ 30,072 | 231 | 219 | 271 | 260 | 245 | Bush |
| Alaska | $ 55,412 | 233 | 212 | 279 | 256 | 245 | Bush |
| Oklahoma | $ 35,500 | 229 | 214 | 272 | 262 | 244 | Bush |
| Georgia | $ 43,316 | 230 | 214 | 270 | 258 | 243 | Bush |
| Arkansas | $ 32,423 | 229 | 214 | 266 | 258 | 242 | Bush |
| Tennessee | $ 36,329 | 228 | 212 | 268 | 258 | 241 | Bush |
| Arizona | $ 41,554 | 229 | 209 | 271 | 255 | 241 | Bush |
| Nevada | $ 46,289 | 228 | 207 | 268 | 252 | 239 | Bush |
| Louisiana | $ 33,312 | 226 | 205 | 266 | 253 | 238 | Bush |
| Alabama | $ 36,771 | 223 | 207 | 262 | 253 | 236 | Bush |
| Mississippi | $ 32,447 | 223 | 205 | 261 | 255 | 236 | Bush |

Source: U.S. Census Bureau, Current Population Survey, 2001, 2002, and 2003 Annual Social and Economic Supplements.

Note: NAEP data were gathered in February 2003. That was the first year that the NCLB "No Child Left Behind" (NCLB) mandates to participate in NAEP took effect—hence it is the first year that all fifty states participated.

posed earlier: how accurate are these curbstoned estimates of IQ? Or at least how much are our conclusions affected by their errors?

There is a substantial correlation (0.66) between the NAEP composite we created and the fake IQ data. Clearly the relationship is not quite as strong as that between the real and fake income data; however, it is considerably larger than the correlation between the NAEP composites and the real income data (.37). But to what extent do the "errors" in the fake data, and the use of a surrogate achievement measure for IQ affect our conclusions?

There are many ways that we might opt to use to compare the fake IQ state scores with the mean NAEP scores. I choose simplicity (there are surely better ways, but I will leave it to others to use them). I have

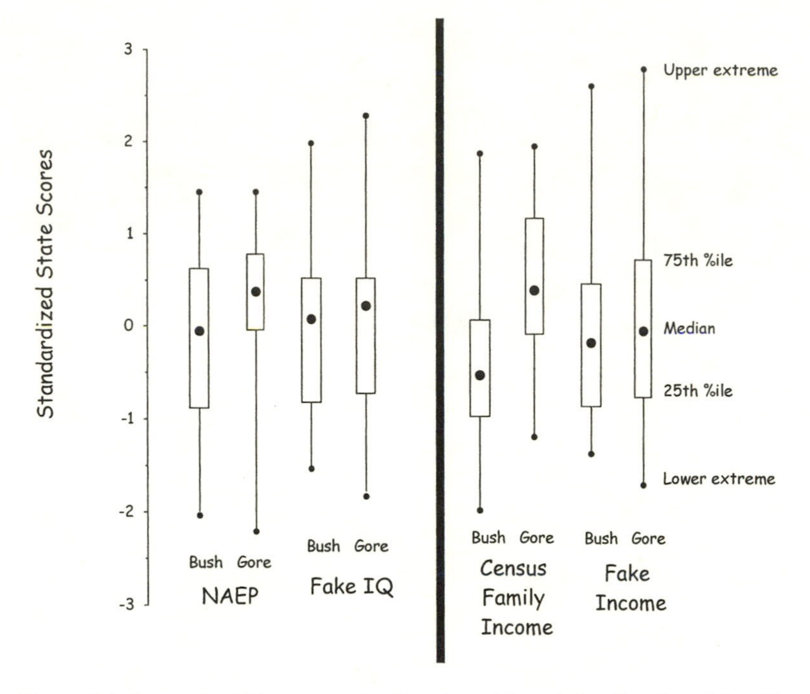


**Figure 2.1.** Box-and-whisker plots showing that, although the direction of the difference between Gore and Bush states was the same for real data as it was for faked, this difference was greater for mean NAEP scores than fake IQ, but smaller for real vs. fake income data.

simply standardized the fake state IQ data in the usual way* and summarized them into box-and-whisker plots† for Gore and Bush states. This is shown for NAEP, IQ, and income data in figure 2.1.

As is now obvious, the effect that was shown with the fake data is qualitatively the same as with the real data; but more extreme for both the test scores and the actual income data. Tukey's observation, reported earlier, on the accuracy of curbstone estimates, seems upheld here. Life often mirrors fantasy, and sometimes is more vivid.

* By subtracting out their unweighted mean and dividing by the unweighted standard deviation—yielding an unadjusted *z* score,

† For those unfamiliar with this type of display, whose purpose is to represent the entire distribution of the data, the box contains the middle 50% of the data with the large dot representing the median. The "whiskers" represent the top and bottom 25% of the data.

3

# Stumbling on the Path toward the Visual Communication of Complexity

Graphical representation has been shown repeatedly over the past two hundred years to be perhaps the best way to communicate complex technical information to an intelligent, lay audience. As with most graphical adventures the tale begins at the cusp of the eighteenth and nineteenth centuries with the remarkable Scot William Playfair (1759–1823). In the introduction to the third edition of his *Atlas*, Playfair (1801, p. xii) tells the readers that if they do not "at first sight, understand the manner of inspecting the Charts, (they need only) read with attention the few lines of directions facing the first Chart, after which they will find all the difficulty entirely vanish, and as much information may be *Obtained in five minutes as would require whole days to imprint on the memory, in a lasting manner, by a table of figures*." (Emphasis Playfair's.)

There have been an enormous number of well-known examples demonstrating the power of graphical displays[1] and so I will include just one here, from the Surgeon General's report on smoking that shows that the increased likelihood of death among smokers is constant across all ages *in a log scale*. The key point here is the use of a log scale. Logarithms, alas, are not part of the everyday intellectual vernacular, despite their being embedded in many of the terms and concepts we use, e.g., the force of earthquakes, the loudness of sounds, the acidity of liquids. Yet, despite their unfamiliar nature, a plot of the

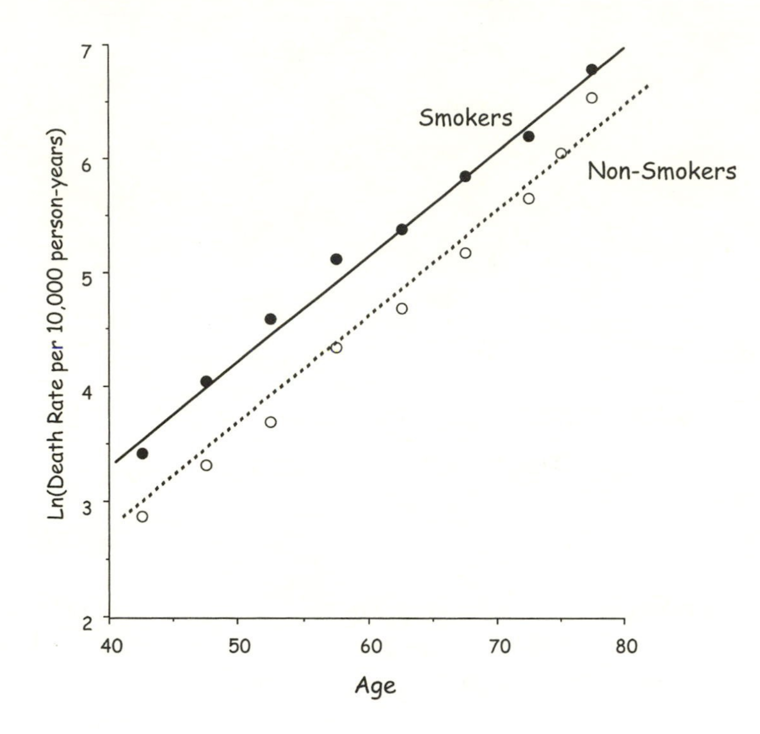


**Figure 3.1.**
Smokers die sooner than nonsmokers. Smoking seems to subtract about seven years from life expectancy.

logarithms of death rates yields a picture with a simple and true interpretation.

Yet sometimes things that appear simple and easy to understand can be misleading, And, ironically, because graphical displays of quantitative phenomena are so easily understood, they can yield memorable misunderstandings. A dramatic example of this appeared on the top of the op-ed page of the *New York Times* on August 4, 2004, in a short (250-word) narrative accompanying two graphs with a very similar structure. Its author was George Shultz, the former Secretary of the Treasury (1972–1974) and Secretary of State (1982–1989). The first of the two graphs showed Gross Domestic Product from 1990 until 2004, the second unemployment for the same time period. These are reproduced here.

The message we get from both of these figures is that the economy dipped during the beginning of the first Bush administration but then turned around; hence Clinton inherited good employment and GDP figures from the first President Bush but turned over an impending

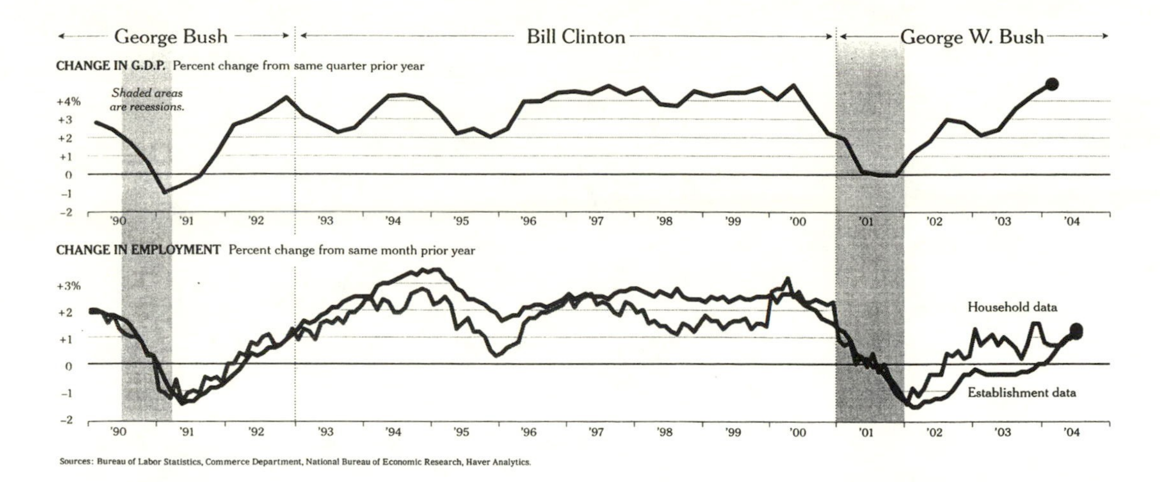


**Figure 3.2.**
George Schultz's specious graphical evidence that "President Clinton inherited prosperity; President Clinton bequeathed recession."

recession to the second Bush who, after about a year, managed to turn it around. This visual impression is reinforced by Shultz's narrative, "President Clinton inherited prosperity; President Clinton bequeathed recession."

But is this true? Consider the lower panel. We note from the plot title that Secretary Shultz did not plot employment rates; he plotted their change. How easily can the lay public correctly understand derivatives? To answer this I turned to the Census Bureau's *Statistical Abstract of the United States*[2] and retrieved the actual employment figures, which I plotted. The result (shown in figure 3.3) tells quite the opposite story. Why was our (yours, mine, and apparently Secretary Shultz's) perception so confused by a plot of the derivative?

Returning to figure 3.2 we note that, during the Clinton years, when the plotted employment curve was level, it meant employment was increasing, and even when it declined at the end of the Clinton administration, the derivative was still positive; hence employment was still increasing, albeit more slowly. In contrast, the derivative was negative at the beginning of the second Bush administration (employment was decreasing) and even when it appeared to turn around, it was still negative for 2002! The same structure holds for the GDP data. Secretary Shultz was not only wrong, but the correct conclusion is exactly the opposite of the one he drew.

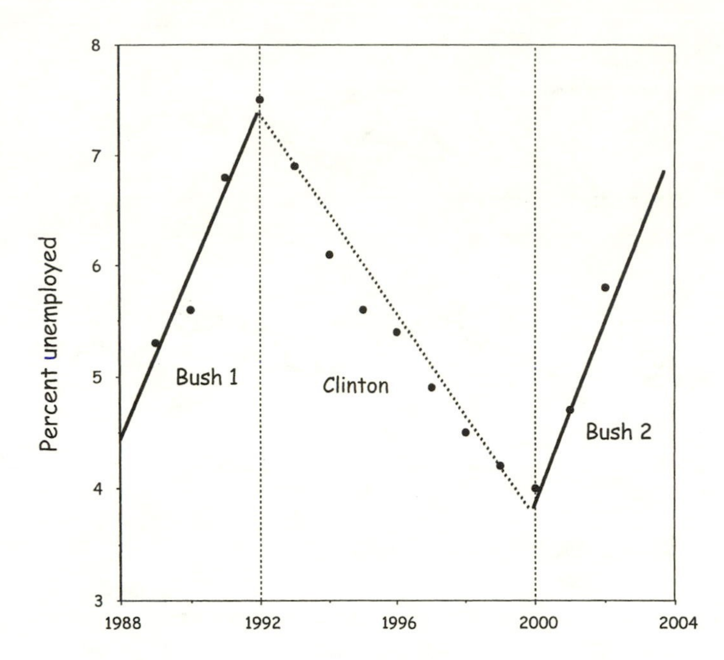


**Figure 3.3.**
Unemployment grew during both Bush administrations but dropped precipitously during the Clinton years.

The remarkable difference between the immediate perception of a plot of the rate of change of employment and the true state of affairs must give us pause. It proves that simply plotting a statistic does not perforce improve its comprehensibility. This example brings to mind Will Roger's well-known aphorism, "What we don't know won't hurt us; it's what we do know that ain't."

4

# Using Graphs to Simplify the Complex: The Medicare Drug Plan as an Example

The goal of graphic display is the simplification of the complex. So it often comes as a surprise to me when this avenue is not exploited more fully by those whose job description includes clear communication. This came to the fore recently over the mangled descriptions of the new Medicare drug plan, in which we, obviously addled-minded, seniors expressed some confusion about the value of the new plan for us, despite the best attempts of the president to explain it.

The source of my misunderstanding was made clear when Robert Pear, a reporter for the *New York Times*, hence a professional explainer, laid out the parameters of the program (June 19, 2005, p. A10):

> Premiums are expected to average $37 a month. Under the standard benefit defined by Congress, the beneficiary will be responsible for a $250 annual deductible, 25% of drug costs from $251 to $2,250 and all of the next $2,850 in drugs costs. Beyond that level—$5,100 a year and more—Medicare will pay about 95 percent of drug costs.

Leaving aside the modifiers "are expected to average" and "about 95 percent," suppose we assume we would pay the average premium and the 95% is exact, should we sign up? This prose describes a situation that reminds me of nothing so much as the rules surrounding tax returns.

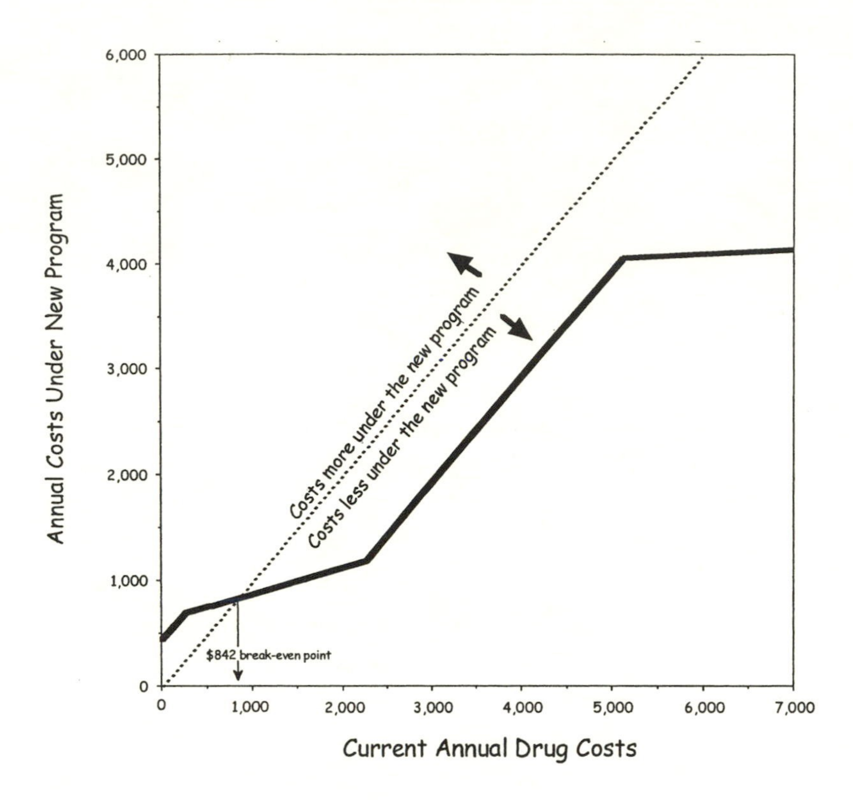


**Figure 4.1.** The new Medicare drug plan can save you money if you spend more than $842 annually on drugs.

On June 17 the president addressed several hundred people at a gathering in the Maple Grove, Minnesota community center: "On the average, the folks who sign up for this prescription drug benefit are going to save $1,300 a year. For the first time in Medicare's history, there will be a stop-loss, a kind of catastrophic care." This sounds great, but we statisticians are always wary of the phrase "on the average," for on the average Bill Gates and I can afford a new Rolls and a winter home in Provence.

I rushed to my computer and made a graph of the drug rules (see figure 4.1). We see that the $37 monthly premium yields an annual cost of $444, which must be paid regardless of how much you use the program. In addition we must pay the first $250 in drug costs. This means that the program costs $694 before any benefits are received. But after this point Medicare starts to pick up 75% of the drug bills and so the benefit starts to eat into the $694 laid out initially. At what

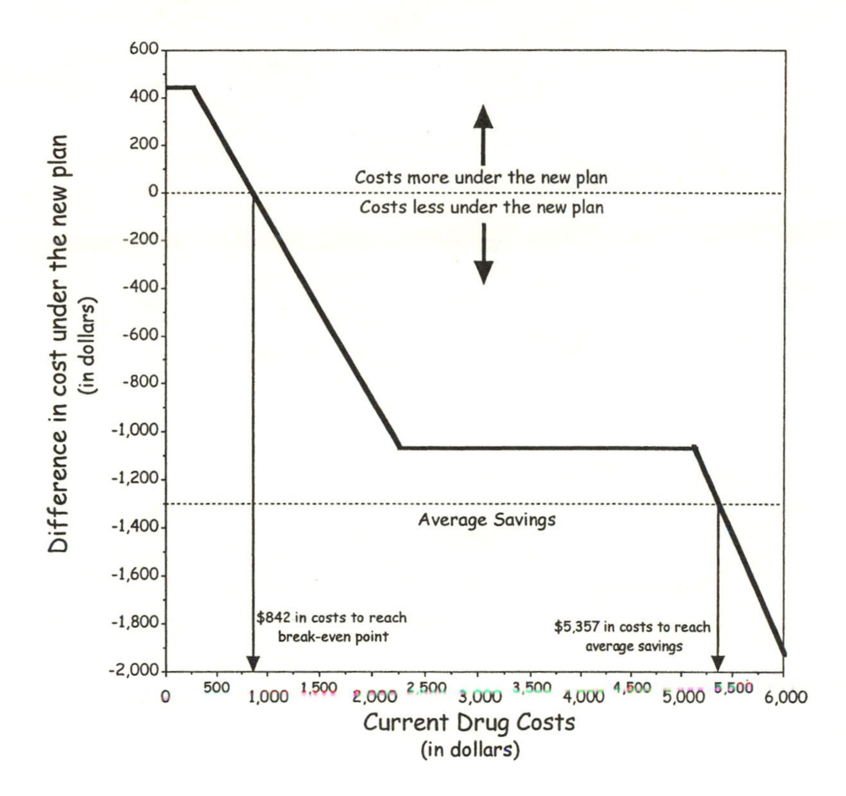


**Figure 4.2.** Graph showing savings under the new Medicare plan as a function of current drug costs.

point do we break even? The graph shows clearly that the break-even point occurs when your annual drug costs equal $842.

Moreover we see that, as our drug expenses continue to rise beyond the $5,100 point, the drug program picks up 95 cents on the dollar. So, for those who have the misfortune to need very expensive drug therapy the president was correct—there is protection against catastrophic drug expenses.

To see more accurately how much the new program can save us, I subtracted the drug costs under Medicare from current costs (see figure 4.2). When the curve is positive, we will pay more under Medicare; when negative it shows how much Medicare is saving us. As we saw before, the savings do not begin until after we have $842 in annual medical expenses, and they remain modest until we pass the $5,100 catastrophe barrier, at which point the real protection begins. The horizontal line at $1,300 represents what the president tells us will be the average saving. This corresponds to an annual drug expenditure of $5,356.84. I don't know what data were used to estimate this figure, but I assume that, if it is accurate, we are seeing the Bill Gates effect—that most of us will see very small benefits of the program but some people will get gigantic ones.

My conclusion about the plan is a positive one, not apparent from the initial description: it works as good insurance ought to, by spreading the risk around. Obviously, it is too expensive to make drugs free to everyone, but this plan does allow the relatively few seniors who require very expensive drug therapy to get it without driving their families into penury. For the rest, who are fortunate enough to not need such therapy, the smallish benefits are likely to ameliorate its modest costs. And even for those whose annual drug bills do not pass the $842 threshold, they may still take comfort in the protection afforded. This conclusion was quite different from my thoughts before I drew the graph.

Two mysteries remain. The first is why have none of the communication pros in the media drawn these graphs? The second arises from multiplying $1,300 by the number of seniors who sign up. This is the annual cost of the program. Where is this money going to come from?

5

# A Political Statistic

> I gather, young man, that you wish to be a Member of Parliament. The first lesson that you must learn is, when I call for statistics about the rate of infant mortality, what I want is proof that fewer babies died when I was Prime Minister than when anyone else was Prime Minister. That is a political statistic.
>
> —Winston Churchill (1874–1965)

On the first page of the U.S. Department of Education website[1] stands a graph, bare of description except for two accompanying quotes: "When it comes to the education of our children . . . failure is not an option" (George W. Bush, U.S. President) and "Every child can learn, and we mean it . . . , excuses are not good enough, we need results." (Rod Paige, U.S. Secretary of Education). That graph (figure 5.1) shows a sharp increase over the past twenty years in the amount spent on education K–12 by the federal government but little change in fourth grade children's reading ability, with 68% of these children still reading at a level deemed less than proficient. Presumably, this graphic is intended to communicate two messages: First, that our children continue to achieve poorly and are not making progress, and second, that spending money is not the answer.

This graphic, however, has four flaws: (1) the left-hand axis shows the wrong measure; (2) the right-hand axis uses an inappropriate scale; (3) the impression given hinges on the subject matter chosen; and

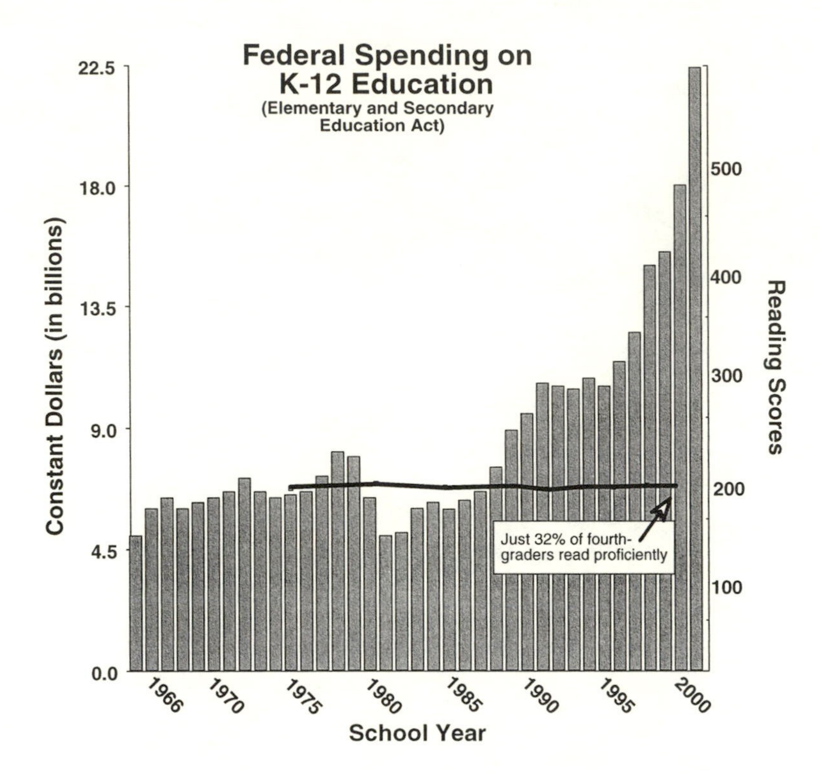


Figure 5.1.
Flawed display purporting to show that money spent on education does not yield positive results (from http://www.ed.gov/, accessed March 23, 2002).

(4) the graphic format is unreasonable. I will briefly comment on all four. In addition, there is of course the problem of omitted variables: other relevant factors, such as the composition of the student population, which did not stay constant during the period spanned by the graphic. I will leave that issue for a future account.

The expenditures shown in figure 5.1 are total federal expenditures in constant dollars. This appears sensible at first blush but is unreasonable in several respects. First, it considers only federal expenditures, which constitute only a small percentage of total K–12 expenditures. Second, it fails to consider the sizable growth in the number of the U. S. student population (figure 5.2), for example, by displaying constant dollars spent per child enrolled. Third, if the graph is to support inferences about cause, one would want reading scores to lag expenditures by some amount, but figure 5.1 continues the expenditure series for two years beyond the reading series—and those two years account for a sizable portion of the total increase in expenditures over the two decades.

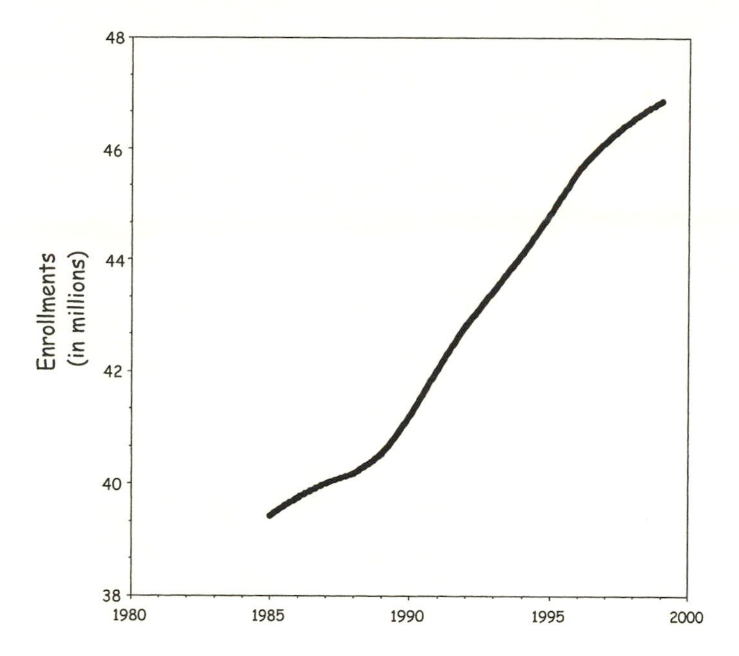


**Figure 5.2.**
Table 40.—Enrollment in public elementary and secondary schools, by grade: Fall 1985 to fall 1999. (From *Digest of Education Statistics, 2001*. Washington: National Center for Education Statistics. http://www.nces.ed.gov//pubs2002/digest2001/tables/dt040.asp, accessed March 24, 2003.)

But what about the right axis? Although it isn't explicitly stated, the reading scores reported are surely those from the long-term trend assessment of the National Assessment of Educational Progress (NAEP). The NAEP scale for reading and writing is one that has a mean of 250 and a standard deviation of 50 across all three age/grade groups in NAEP's entire student sample, which includes fourth, eighth, and twelfth grade students. In the last year displayed (1999), the range of performance in the fourth grade (actually, the age-9 sample) from the fifth to the ninetyfifth percentile was just under 130 points. Displaying the reading trends on a scale almost four times that large preordains an appearance of stasis even when moderately sizable improvements are taking place.

If we reset the scale to more practical bounds, and show the expenditures on a per pupil basis, we see (figure 5.3) that, while there have been linear increases in per pupil expenditures, the changes in fourth grade reading scores have been more erratic. But, if we discount the peculiarly high score in 1980, there is some evidence of growth over the quarter century shown, particularly if one includes the earliest data point (1971), which does not appear in the graphic. My point is

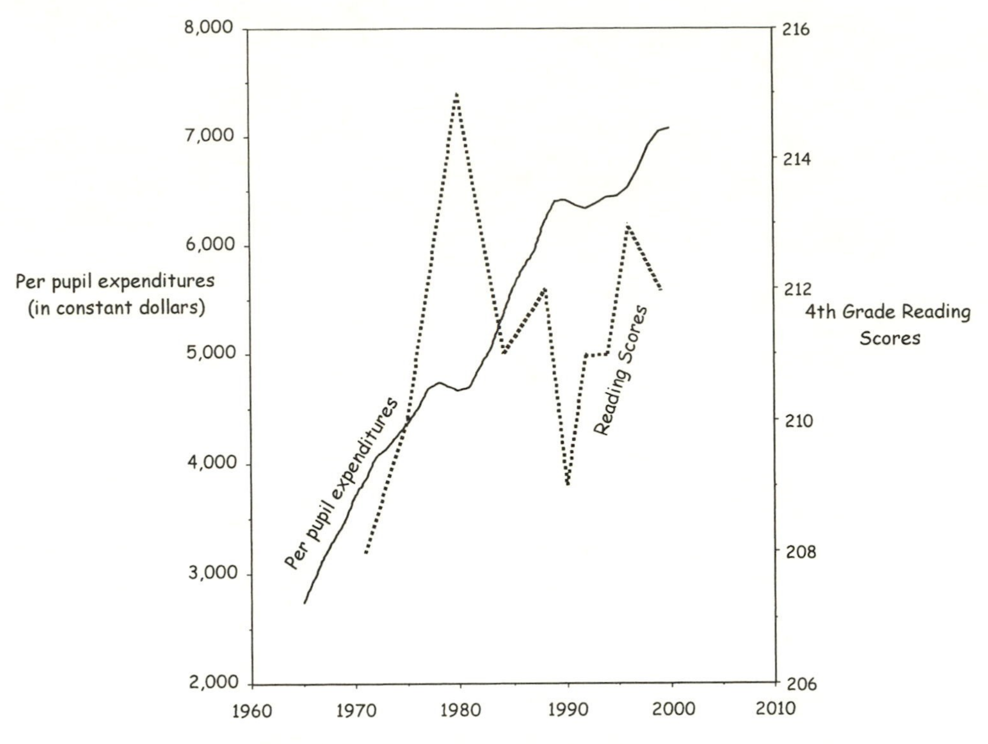


**Figure 5.3.** Per pupil expenditures and reading scores. Per pupil expenditures are taken from Table 167, Total and current expenditure per pupil in public elementary and secondary schools: 1919–20 to 2000–01. *Digest of Education Statistics, 2001*. Washington: National Center for Education Statistics. http://www.nces.ed.gov//pubs2002/digest2001/tables/dt167.asp, accessed March 24, 2003. Reading scores are taken from the long-term trend assessment of the National Assessment of Educational Progress. See http://www.nces.ed.gov/nationsreportcard/tables/Ltt1999/, accessed March 23, 2003.

not that the message provided in figure 5.1 is necessarily false, but rather that its form prevents the viewer from drawing the more nuanced inference that is easy from figure 5.3.

So far we have focused on how data are displayed. Next we move to what data were chosen to be displayed. Although fourth grade reading is a critical aspect of education, there are other aspects also worthy of attention. For example, if mathematics was chosen to be shown instead of reading, we would find that there has been substantial improvement

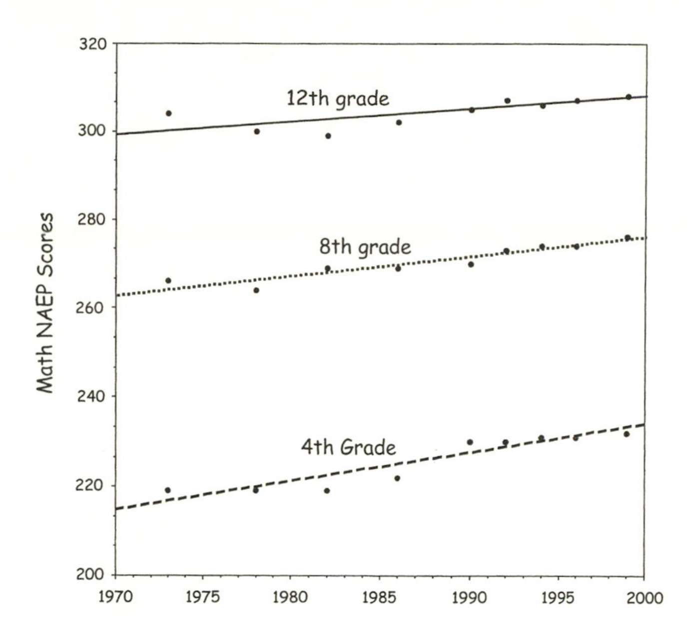


Figure 5.4.
NAEP math scores have shown steady improvement in all grades over the past 25 years. Math trend data are taken from the long-term trend assessment of the National Assessment of Educational Progress. See http://www.nces.ed.gov/nationsreportcard/tables/Ltt1999/, accessed March 23, 2003.

over these years in all grades (see figure 5.4). In fact there is also substantial improvement shown in reading scores for both eighth and twelfth grades. It is only fourth grade reading that does not show clear gains.

I have elected to show the progress in student mathematics performance separately from expenditures, which ought to occupy its own graph with scale chosen independently of the test score scale. It is very difficult to use the double *y*-axis format to present results honestly, and it too often leads to incorrect inferences. The impression given by these graphics is often a matter of arbitrary choices of scale. For example, suppose we show both fourth grade math and per pupil expenditures as a double *y*-axis plot using each axis to maximally spread out its associated data (see figure 5.5). The conclusion shown as the figure caption, that both variables have moved apace, is one that is naturally suggested. But in fact it is merely the outcome of two (roughly) monotonic variables being plotted independently.

We could, just as easily, plot these data in another way (see figure 5.6) that suggests that America's students were progressing well despite the penurious support their schools were receiving from the government.

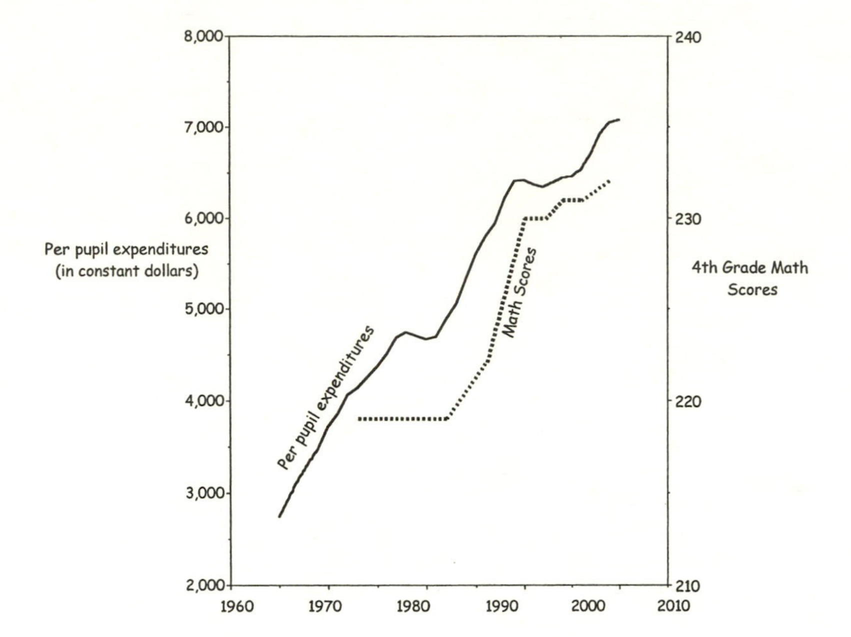


Figure 5.5.
Spending on education and fourth graders' performance on math have improved apace. See source notes for figures 5.3 and 5.4.

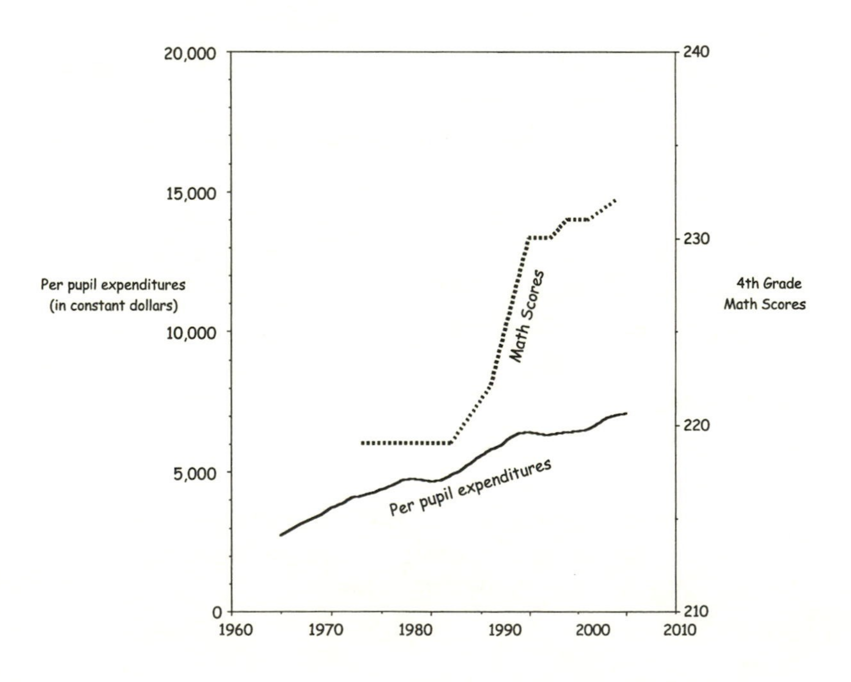


Figure 5.6.
Fourth graders' performance in math has improved despite lagging funding for education. See source notes for figures 5.3 and 5.4.

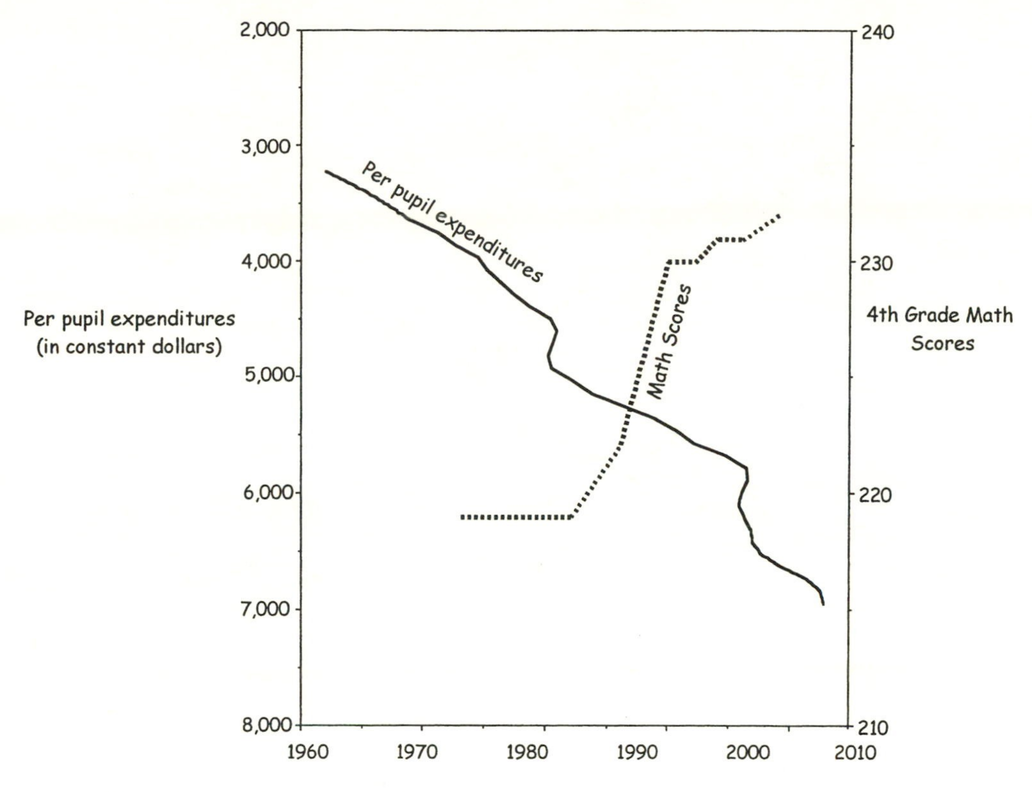


**Figure 5.7.** Fourth graders' performance in math has continued to improve despite education funding. See source notes for figures 5.3 and 5.4.

The double *y*-axis graph allows the plotter (in the pejorative sense as well) to play any game at all, making the variables shown appear to have whatever relationship is desired. Even if one variable goes up while the other one goes down it does not present a serious obstacle to this remarkably flexible format. To change the meaning all you have to do is reverse one axis. So, if we puckishly use this display format's power more fully, we can completely reverse the impression given initially (see figure 5.7).

Those who believe Pollyannishly that no one would actually try to get away with such tomfoolery need look no further than the plot shown as figure 5.8 (accurately redrafted from the *Ithacan*, October 18, 2000) to see that this can be done sufficiently subtly so as to lead even careful readers astray. It looks as if Cornell tuition keeps rising despite

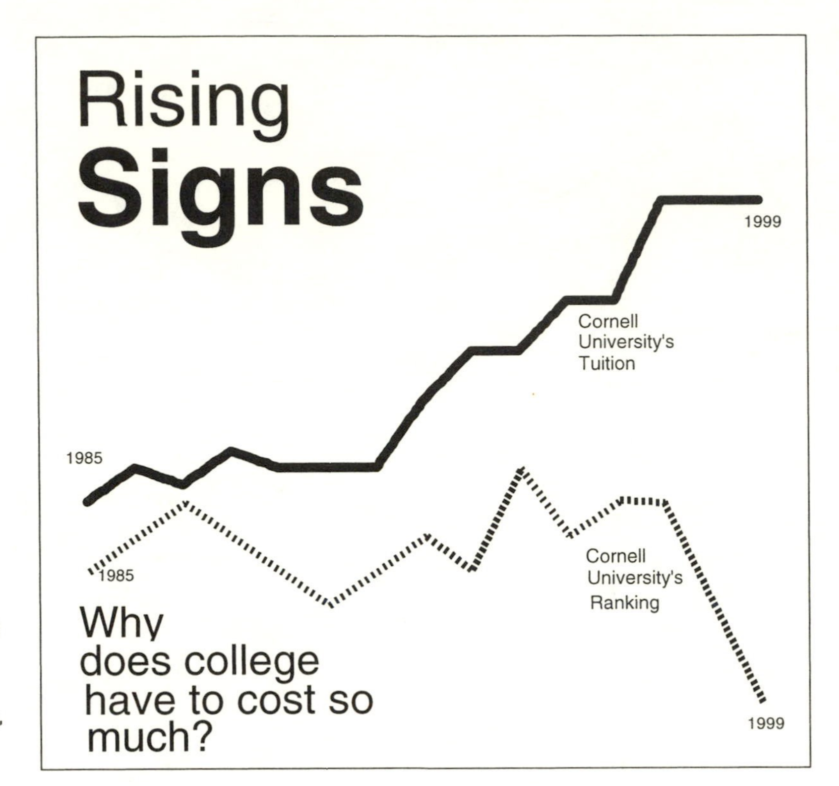


**Figure 5.8.**
Double *Y*-axis graph that reverses the direction of one axis to mislead the viewer (from The *Ithacan*, October 18, 2000, p. 1).

the nosedive taken by its ranking. It is only after we realize that the leading university has a rank of one that we see that a dropping rank is a good thing!

This exercise in the critical examination of a government plot reminds us of the wisdom of Mark Twain, who, like Churchill after him, advised that we should "Get your facts first, and then you can distort them as much as you please. (Facts are stubborn, but statistics are more pliable.)." Remember George Schultz's oft-quoted epigram, "The devil is in the details."

6

# A Catch-22 in Assigning Primary Delegates

As the 2008 election loomed ever closer, many states maneuvered in various ways to try to gain increased influence. The goal for New York, as a March 31, 2007, editorial in the *New York Times* put it, should be to (i) "have a chance to play a major role in choosing their party's nominee," and (ii) "make the candidates address New York's issues." The editorial continues, "The candidates always rave about ethanol in Iowa. If they had to hunt for votes in the Bronx and on Long Island, they might talk more about mass transportation and middle-income housing."

The principal action taken by New York's legislature to accomplish these goals was to move the date of the primary election forward from March to February. In the past, much of the political dust had already settled by the time New York had its primary, and so there were few practical consequences except to second what had already been decided by the earlier states. 2008 was different as the biggest states moved their primaries to February, setting up a "Super Tuesday" when it was thought that the candidates for the two major parties would be determined.

But would merely moving the primary date forward be sufficient? The *Times* did not think so. The editorial argues,

> It will still be an easy state for campaigns to skip. For starters, there are already home-team favorites — Senator Hillary Clinton

> for the Democrats and former Mayor Rudolph Giuliani for Republicans.
>
> The more entrenched problem is New York's rules. In the Republican primary, the winner of the popular vote gets all the delegates. If Senator John McCain gets, say, 40 percent to Mr. Giuliani's 50 percent, he would win zero delegates. Given how expensive it is to run in New York, many candidates will probably not bother.
>
> The Democrats' selection process is better but not much. Candidates are awarded delegates proportionately, but by Congressional districts.

If awarding delegates in an all-or-none manner, whether for the state as a whole (Republicans) or by congressional district (Democrats), is not going to aid in attaining the goals of giving New York its fair share of influence, what would be better? The *Times* suggests,

> In New York, Gov. Eliot Spitzer, the Legislature and the parties should fix the state's arcane rules. Ballot access should be simplified, and delegates should be awarded based on shares of the statewide vote.

The democratic notion of proportional representation sounds good, but will it work? To get a little ahead of ourselves, maybe not. But that sort-of conclusion awaits a more careful examination, which we explore using a small example[1]. It is this example that provides us with the insights that form the core of our conclusions.

### 6.1. THE EXAMPLE OF NINE VOTERS

Suppose we are one of just nine voters and we would like to maximize the chance of our vote making a difference. What should we do? Let us begin with a definition of what it means to have our vote count. In this instance, let us say that our vote counts if we cast the deciding vote. So, if the other eight voters are divided eight to zero, it doesn't matter how we vote. In fact it will only matter if the other eight are divided four to four. What is the probability of this happening? Let's work with the simplest statistical model and assume that all combinations of votes are equally likely, that is, each voter has a 50% chance of choosing each candidate. Then the probability of the eight other voters dividing four to four is simply

$$(½)^8 \times \text{the number of ways that 4 votes can be taken from 8}$$

or

$$(1/256) \times (8 \times 7 \times 6 \times 5)/(4 \times 3 \times 2) = 70/256 = 0.273. \quad (6.1)$$

Is there a way to increase this? Suppose we get together with four other voters and form a bloc. We agree that we will vote among ourselves first, and however the majority decides is the way we will cast all five votes. By forming this bloc we make the remaining four voters irrelevant. But what is the chance of our vote counting now? Simply put, our vote will count if the other four voters in our bloc divide two to two. The probability of this happening is

$$(½)^4 \times \text{the number of ways that 2 votes can be taken from 4}$$

or

$$(1/16) \times (4 \times 3/2) = 12/32 = 0.375. \quad (6.2)$$

This simple example has provided the insight to see why voting as a bloc increases our influence. Thinking back to the allocation of delegates, we can see that, if any state decides to allocate their delegates proportional to the popular vote, that state will yield an advantage of perhaps only one or two delegates to the winner over the loser. If it is an expensive state to campaign in (such as California or New York), a sensible strategy would be to skip that state and spend maximal effort in an all-or-none state where the potential rewards are far greater.

Is this all? No. Our nine-vote example has further riches that have yet to be mined. Suppose we cannot get four other voters to form a bloc, and instead all we could get was three. What are the chances of our vote counting now?

This calculation has two parts. The first part is the chance of our vote counting in our bloc. That will happen if the other two voters split one to one, which will happen one-half of the time. The second part is the probability of the other six votes splitting in such a way that our three-vote bloc is decisive. That will happen if those other six votes split four to two, three to three, or two to four.

The chances of this happening are substantial, to wit,

$$(½)^6\,(6 \times 5 + 5 \times 4) = (50/64);\ \text{hence the probability of our vote counting is}$$
$$(1/2) \times (50/64) = 50/128 = 0.391. \tag{6.3}$$

This strategy yields the greatest chance of our vote counting of the three cases we have examined, and would suggest that perhaps forming less-than-certain blocs is the path toward optimization. But this sort of thinking is a snare and a delusion. Let us make the assumption that the other six voters are just as clever as we are and they decide to form blocs as well. If they form a bloc of five or six. our vote will not count at all, but suppose they imitate us and form themselves into two blocs of three. Now our vote will count only when those two blocs split. The probability of such an occurrence is one-half, and so the probability of our vote counting is the product of it counting within the bloc (½) and the other two blocs splitting (½), yielding a probability of 0.25.

Compare this figure with the 0.273 given in expression (6.1) for the probability of our vote counting in an unaligned situation. This is directly analogous to the so-called Prisoner's Dilemma (see sidebar) where, if we don't form blocs and some others do, the power of our vote is diminished. But if we all form blocs, everyone's influence is less. This is the Catch-22 referred to in the title.

### 6.2. IMPLICATIONS FOR NEW YORK—AND OTHER STATES

My argument suggests that, on average, voters might gain if all states were to adopt rules that gave proportional representation based on the primary vote—but if even one state demurred and opted to form a bloc by using an all-or-none allocation scheme, it is likely that every state but that one would lose influence. This in turn would motivate each of the other states to switch to an all-or-nothing allocation to increase their own probabilities of influencing the outcome. (Blocs can also be formed by groups of states, making our example more complex by adding nested coalitions; in a sense, this is what occurs when a group of states band together to create a Super Tuesday of primaries on a single day, with the goal that the winner of this group of primaries will have a better chance of gaining a decisive number of delegates.)

However, these calculations depend on the assumption that voters are closely divided as to which candidate to support. To study the implications of allocation rules for actual elections, we need to go the next step and actually estimate the probability distributions using historical votes. A key factor is the closeness and anticipated closeness of the elections within each state[2]. In particular, if the Democrats (say) think that Hillary Clinton is almost certain to win New York and all its delegates, then they might want to consider the *Times* suggestion and switch to a proportional system where all votes will be counted—

**Sidebar for the Prisoner's Dilemma**

In the 1940s and 1950s the mathematical structure of game theory was being developed and exploited. John Nash's path-breaking work in game theory is what resulted in his 1994 Nobel Prize. A signal point in game theory was a 1950 lecture by the Princeton mathematician Albert W. Tucker who first proposed what has come to be called the Prisoner's Dilemma. In its simplest terms consider two suspects arrested after a robbery. They are held in two separate rooms and are unable to communicate. The district attorney approaches each of them separately with a deal. "If you confess, and agree to testify against the other, you will get a one-year sentence and the other person will get ten years. If you both confess and cooperate you both will get five years. And, if you both remain silent, you both will get a two-year sentence."

Analyzing this punishment system carefully, we can see that, if the other person stays silent, your best option is to cooperate. But if the other person cooperates, your best option remains to cooperate as well. Thus, no matter what the other person does, your best strategy is to cooperate.

Assuming that your accomplice is every bit as clever as you are, the same logic will apply, leading both of you to cooperate and both to end up with five-year sentences.

Yet, if you both had remained silent, your sentences would have been just two years. The dilemma is obvious—it is hard to arrive at the winning strategy independently, for even if you agree in advance to remain silent, if the other one does not, he is better off and you are punished. And vice versa.

although this would have the side effect of decreasing the delegate total for hometown favorite Hillary. But if there is a good chance that the primary election will be close, New Yorkers would actually have more influence with a winner-take-all system.

If New York Republicans (or Democrats) prefer a proportional system for allocating delegates, the best way to accomplish it might be to work in concert with other states so as to avoid the Prisoner's Dilemma problem that allows any state to gain by switching to a winner-take-all system. They could take a cue from the state of Maryland, which, as reported by Brian Witte in the Associated Press on April 10, 2007,

> officially became the first state on Tuesday to approve a plan to give its electoral votes for president to the winner of the national popular vote instead of the candidate chosen by state voters. . . . The measure would award Maryland's 10 electoral votes to the national popular vote winner. The plan would only take effect if states representing a majority of the nation's 538 electoral votes decided to make the same change.

To beat a bloc, form a bloc of nonblockers.*

*As this chapter was being written, the all-or-nothing character of the Republican Republican primaries has resulted in a nominee. At the same time, the proportional representation yielded by the Democratic primaries has meant that neither of the two Democratic front-runners has yet achieved the required number of delegates to assure the nomination. Instead they continue to expend resources, both financial and emotional, beating on one another for the small advantage in delegates that each primary offers. Since the ultimate goal of the primaries is to select a candidate with the best chance of winning in the general election, we will not know which approach was superior until the election. Even then we will not be sure, for we will not be able to know what the results would have been otherwise. But some evidence is better than none – so long as we know of the shortcomings.

# PART III Educational Testing

In most branches of civilized society, decisions must be made on the allocation of resources and labor. In primitive societies this might have involved resolving issues of who will go afield to hunt and who will remain behind to see after the home and children. In more recent times, this approach can be thought of as a metaphor for going to work and staying home. In a rational world the decision as to who would do what would be made on the basis of who was best able to accomplish the tasks. Of course, unless one actually tries out both roles, this decision will be made with uncertainty. For a long time the decision was made more or less solely on the basis of sex. As we entered the industrial age, various jobs were staffed on the basis of many different criteria, too often having to do with sex or social class rather than through some rational system of triage. In western society the notion of using proficiency testing to reduce the uncertainty in making such decisions was of relatively recent origins (beginning to be widely used only in the nineteenth century, and even then the decisions were still dominated by sex and social class). Yet the use of testing to help reduce the uncertainty in the allocation of resources is indeed ancient.

Some rudimentary proficiency testing took place in China around 2200 B.C., when the emperor was said to have examined his officials every third year. This set a precedent for periodic exams that was to persist in China for a very long time. In 1115 B.C., at the beginning of the Chan Dynasty, formal testing procedures were instituted for candidates for office. Job sample tests were used, with proficiency required in archery, arithmetic, horsemanship, music, writing, and skill in the rites and ceremonies of public and social life.

The Chinese testing program was augmented and modified many times through the years and has been praised by many western scholars. Voltaire and Quesnay advocated its use in France, where it was adopted in 1791, only to be (temporarily) abolished by Napoleon. British reformers cited it as their model for the system set up in 1833 to select trainees for the Indian civil service—the precursor to the British civil service. The success of the British system influenced Senator Charles Sumner and Representative Thomas Jenckes in the development of the civil service examination system they introduced in Congress in 1868.

In the four thousand years since its inception, mental testing has promised to provide the path toward a true meritocratic society. Replacing family connections with an individual's ability as the key to open the doors to economic and social success remains a principal goal of modern societies. Progress toward this goal has been impressive, but it has occurred in fits and starts. In this section we examine three proposals to aid in using test scores toward making this a more just society. The first (chapter 7) uses a statistical method commonly employed in other circumstances to solve a vexing problem. In chapter 8 we examine a well-meaning, but, at its heart, flawed scheme, aimed at reducing intergroup differences. And finally in chapter 9 we look at a recent court case involving test scoring and show that the defense's case was based on a misunderstanding of the meaning of uncertainty.

# 7

# Testing the Disabled: Using Statistics to Navigate between the Scylla of Standards and the Charybdis of Court Decisions*

Either/or is the path to salvation; and/both is the road to hell.

—Kierkegaard (1986, p. 24)

Kierkegaard illustrates the frustration we often face in making difficult decisions with a long list of situations in which many would like to choose both of two incompatible options ("If you have children, you will regret it. If you don't have children, you will regret it. Whether you have children or not, you will regret either."). His characterization of the situation when we are compelled to adopt both (the road to hell) seems accurate. Such situations arise, alas, too frequently, and when they do, we must find ways to comply. One such situation arises in the testing of disabled individuals.

The problem begins innocently enough with the 1985 *Standards for*

**Scylla* is a large rock on the Italian side of the Straits of Messina facing *Charybdis*, a dangerous whirlpool. I use the terms allusively to describe the perils of running into a new evil when seeking to avoid its opposite.

*Testing* that were prepared by a blue ribbon committee of the American Psychological Association. It states (Standard 1.2) that

> if validity for some common interpretations has not been investigated, that fact should be made clear, and potential users should be cautioned about making such interpretations.

This is made more specific in the chapter on the testing of disabled persons when (Standard 14.2) directs

> Until tests have been validated for people who have specific handicapping conditions, test publishers should issue cautionary statements in manuals and elsewhere regarding confidence in interpretations based on such scores.

There are many kinds of accommodations that are made in the course of testing disabled persons (e.g., test forms in Braille or large type, making available an amanuensis to read the questions and write down the answers), but the most common accommodation made, by far, is increasing the amount of time allowed for the test.[1] This modification is often done because it seems sensible. Historically there have been little or no validity data gathered on tests when administered in this way. Indeed, in a 2000 summary of five studies that were run to test the fairness of this strategy it was concluded[2] that "extending extra examination time to learning disabled students appears to violate one . . . criterion for test integrity." The major reason for this lack of data is the relatively small number of disabled test takers of any particular character. Again, the test standards point out, "For example, there are usually not enough students with handicapping conditions entering one school in any given year to conduct the type of validation study that is usually conducted for college admissions tests."

To satisfy Standards 1.2 and 14.2, testing companies typically used to identify tests that are not given under standard conditions. This "flagging" describes the conditions under which the test was administered and advises the users of the test scores that the claims made for the test may be weaker for these scores.

On November 1, 1999, Judge M. Faith Angell of the U.S. District Court for the Eastern District of Pennsylvania issued an injunction prohibiting the National Board of Medical Examiners (NBME) from flagging scores earned under nonstandard testing conditions on the U.S. Medical Licensing Examination. However, the injunction was temporarily suspended pending further briefing at the Appeals Court level by chief judge Edward Becker. The plaintiff, a fourth-year medical student called "John Doe," filed the lawsuit under Title III of the Americans with Disabilities Act (ADA), which prohibits places of "public accommodation" from discriminating on the basis of disability. Doe has multiple sclerosis and received 1.5 times the normal testing time and a seat near the bathroom. He argued that the practice of flagging scores of tests taken under nonstandard conditions unnecessarily identifies him as a person with a disability, thus unfairly calling into question the validity of his scores and subjecting him to discrimination based on the disability. The ruling noted that the practice of flagging test scores could lead administrators of medical residency programs to discriminate against a medical student who receives special accommodations for a disability. The NBME appealed the decision to the U.S. Court of Appeals for the Third Circuit in Philadelphia. NBME lawyers argued that the ADA could not be used against the NBME to bar discrimination that could occur by third parties, such as hospital residency programs. The Appeals Court has ordered a further briefing on the case before deciding whether or not to overturn the District Court ruling. Knowledgeable observers agree that this is a bellwether case on the issue of flagging.

To sum up, testing companies must identify tests administered under conditions in which they cannot vouch for the validity of the usual inferences. They must also provide special accommodations for disabled persons to allow them to participate fairly in the testing. Moreover, there are insufficient individuals with any given disability to allow credible validity studies, yet they cannot identify individual test scores that were obtained under nonstandard conditions. Scylla and Charybdis indeed!

Let me offer one approach to this problem* which I will call "asymptotic score." This approach is a lineal descendent of the method of low-dose extrapolation (and, for that matter, accelerated life testing) that is

* This method is surely not original with me. I remember Don Rubin in a conference discussion in 1986 making a suggestion very much like this. I assume there have been others as well.

widely used in biostatistics. Low-dose extrapolation is used in so-called "Delaney Clause" research that tries to estimate the carcinogenic effect of various substances. Experimenters might give one group of rats the equivalent of twenty cases of diet soda a day and note the number of tumors that result. Then they would give a second group of rats the equivalent of ten cases a day and a third group one case. They then plot the number of tumors as a function of the amount of diet soda and extrapolate down to what a normal daily dosage would be. This is then used to predict the carcinogenic effect of the sweetener in diet soda.*

Accelerated Life Testing is a parallel procedure. When I read an advertisement for a long-life light bulb telling us that it will last 50,000 hours, I invariably wonder how they know this. Did they actually delay marketing the bulb for the 5.7 years it would take to establish this life span? I doubt it. Instead, they probably burned one set of bulbs at 200 volts, another at 190, a third at 180, etc., while noting how long it was before the bulbs burned out. The connective function is then extrapolated downward to predict life span when the voltage is 110.

Of course there are many statistical and logical problems with this approach, but it is a plausible methodology when the more appropriate experiment is too time consuming or too expensive to run.

How can this methodology help us to derive fair test scores for disabled people? It seems sensible to construct an empirical function that connects the amount of time given to take the test with the score obtained. To construct such a function, we might divide up a set of nondisabled people randomly and have one group take the test in one hour, a second group in two hours, a third in three hours, and so on. Then we keep track of the mean score for each group. A smooth curve fitted to such data might look like that shown in figure 7.1. The hypothetical data show that those given an hour for the test would score 385 on average; whereas, if they were given an unlimited amount of time, they would, on average, score 600.

This score-by-time function can be used to adjust the scores of test takers who were given differing amounts of time. Obviously the adjustment can be to any specific time, so we can say that someone who

* They also extrapolate from rats to humans, which may be an even larger jump than from one case a day to two cans.

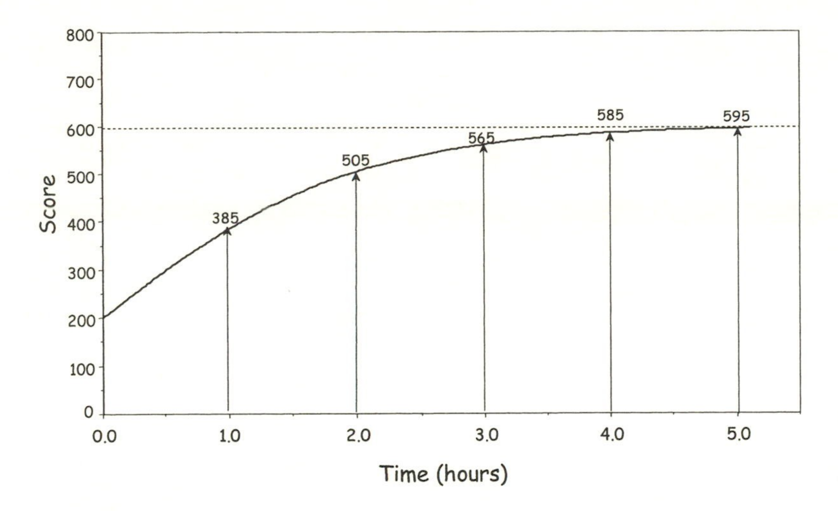


**Figure 7.1.** Hypothetical time-score curve.

got 385 if only allotted an hour would get (on average) 600 if given an unlimited time; or vice versa. This is the key idea in making the scores of disabled and nondisabled people comparable and so provide a credible basis for removing the "flag." Thus my notion is this. Give each test taker two scores: one is the score they obtained, along with a description of the circumstances under which it was achieved, and the second is their asymptotic score, the adjusted score they would have obtained had they had an unlimited amount of time.

Of course the validity of this adjustment rests on an assumption that the test is testing the same thing when it is given with an unlimited time as it is when it is (at least partially) speeded. But this assumption underlies current practice to some extent as well. It certainly underlies the recent decision of the U.S. District Court for the Eastern District of Pennsylvania. Since disabled test takers who take the test under the condition of unlimited time will not be subject to any adjustment, it makes no assumption about the comparability of the score-by-time functions for these two groups.

The method of gathering data deserves further discussion. Note that examinees were randomly assigned to the various timing situations. This requires a special data gathering effort. It would not be fruitful to simply give examinees all the time they want and keep track of how long they took, for in this latter circumstance the amount of time

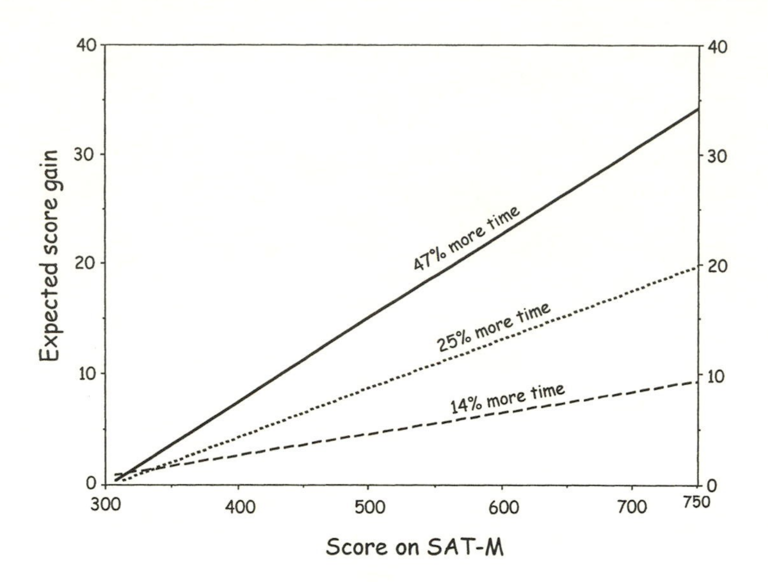


**Figure 7.2.**
The more time available on SAT-M the greater the gain. Expected gain scores on the various experimental math sections over what was expected from the standard length section. Results are shown conditional on the score from the appropriate operational SAT-Math form.

is not an experimental treatment but rather an outcome, and as such is not under experimental control. What we would be likely to observe is that those with the highest scores also tended to take the least amount of time. If we were to believe this, we would recommend that the pathway toward higher scores is giving examinees less time to complete the test. An obviously dopey conclusion, but surprisingly not universally recognized as such—see Freedle's Folly, discussed in chapter 8.

It is operationally difficult to use differential timing, so in an experiment done using a randomly chosen subset of 100,000 examinees from the October 2000 administration of the SAT, we kept the time of a section constant, but varied the number of items presented in that time period. We then scored only those items that all experimental groups had in common and found that on the SAT-Mathematics test, more time yielded higher scores, but on the SAT-Verbal test, the extra time had no effect. But the results provided a richer story yet.

In figure 7.2 we see that, while extra time did indeed yield higher scores, how much higher depended on the individual examinee's ability, with higher-ability examinees gaining more from the extra time than lower-ability examinees. Very high-ability students with scores near 750 on the normally timed exam gained, on average, almost 35

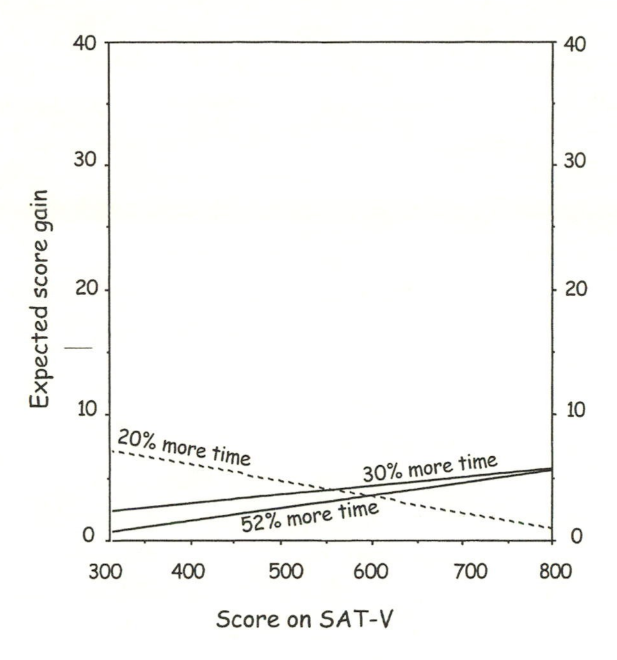


**Figure 7.3.**
Extra time on the verbal section has a small and inconsistent effect. Expected gain scores on the various experimental verbal sections over what was expected from the standard length section. Results are shown conditional on the score from the appropriate operational SAT-Verbal form.

points; whereas examinees with scores near 300 garnered no benefit from extra time. If these results are demonstrated to hold up on further replication, they would predict that group differences in mean SAT-M scores would be exacerbated if the time to take the test were to be increased.

The mathematics results contrast sharply with what was observed with data from the SAT-Verbal section (see figure 7.3), where extra time had little or no consistent effect regardless of the examinee's score.

It is remarkable how often, after we have data, the newly obtained results set off a light in our brains and we realize that the results could not have been any other way. In this instance we see that, in math, if you are completely at sea, having extra time yields nothing but increased sea-sickness. But if we understand the material, just a little extra time may be all that we need to work out the answer. On a verbal test, however, in which vocabulary is often a critical component, if you don't know the words, extra time will not help.

Regardless of the fine structure, however, it is clear from these results that there are plausible conditions where giving extra time to some examinees and not to others can convey an advantage. Thus, when allocating extra time to an examinee with a specific disability, one must give enough to make up for the effects of that disability but not too much so as to convey an unfair advantage. Determination of that amount requires experimentation that may be beyond practical bounds. The insight provided by the experimental study that varied time adds support to the notion that something needs to be done, and giving unlimited time to examinees with disabilities while adjusting all other examinees' scores to what they would have been with unlimited time may be a practical solution.

But this is not what was done. Since these court decisions, many testing companies have decided to avoid further litigation by continuing to make accommodations for examinees with disabilities, but, in apparent violation of Standard 14.2, have ceased flagging scores obtained under nonstandard conditions. Since 2000 both the Educational Testing Service and the College Board have agreed to discontinue the practice of flagging test scores obtained under nonstandard conditions. Thus users of test scores provided by those organizations have no choice but to treat all scores as equivalent, regardless of the conditions under which they were obtained.

This practice has led to a substantial increase in applications for extra time and a concomitant increase in the scrutiny with which such applications are greeted.

8

# Ethnic Bias or Statistical Artifact? Freedle's Folly

The child in me was delighted.
The adult in me was skeptical.

—Saul Bellow, 1976, upon receiving the Nobel Prize for literature

## 8.1. INTRODUCTION

In 1971, the U.S. Supreme Court in *Griggs v Duke Power Company* restricted the use of tests with group differences in scores to those specifically validated for the job in question.[1] In 1972, the National Education Association, noticing the increased vulnerability of tests, called for a moratorium on the use of standardized tests in the nation's schools "because of linguistically or culturally biased standardized tests."[2] This rather extreme point of view vied with Anne Cleary's more scientific concept that "a test is fair if it neither systematically underpredicts nor overpredicts for any group."[3] After more than a decade of debate the 1985 edition of *Standards for Educational and Psychological Tests*[4] cited Cleary's model as "the accepted technical definition of predictive bias." This definition was maintained in the most recent (1999) elaboration of the *Standards.*

Many lay people assume, because of the group differences observed on the SAT, that the test is biased and hence underpredicts the college performance of Black test takers. As is well known now, however, the SAT overpredicts the performance of Black test takers.[5] These findings, combined with the dearth of data supporting the view that large-scale standardized tests show bias against underpriviledged minorities, led to recent Supreme Court decisions (e.g., O'Connor's decision regarding law school admissions: *Grutter v Bollinger*, 123 S. Ct. 2325 [2003], and Rhenquist's decision regarding admission in colleges of literature, science, and arts: *Gratz v Bollinger*, 123 S. Ct. 2411 [2003]) that have made the use of race or sex in college admissions decisions increasingly difficult.*

Although score differences alone are not proof of bias, if the differences are not related to the measured constructs, the validity of tests are diminished by the extent to which the test contains construct irrelevant variance.[6] If the construct irrelevant variance is related to race it yields a race bias.[7] If it is related to sex it yields a sex bias.[8] If it is related to language it yields a linguistic bias.[9] Although there has been an enormous effort to rid the SAT of all construct irrelevant variance, some still remains. Some is due to speededness, because practical constraints preclude extending the test for 50% more time. Some is due to language, because it is hard to ask word problems without words. Some is due to cultural differences, because certain types of questions and scenarios are more familiar for some subcultures than others. Nevertheless, although some bias of these sorts still exists on the SAT, typically its size is smaller than regression effects that push in the opposite direction. Thus, the existing test has a positive bias that favors Black and Hispanic examinees, overpredicting their success in college. There is a slight negative bias against women in mathematics[10]—an underprediction of women's performance in college math courses. Similarly, there is an underprediction of college success for White and Asian examinees.[11]

Such findings have led to a search for acceptable objective measures that would allow the fulfillment of a commonly held societal goal: the fair representation of all demographic groups within highly selective institutions. The present imbalance is in part due to underrepresented minorities typically receiving lower average scores than the majority

* For more details see http://www.umich.edu/news/Releases/2003/Jun03/supremecourt.html

competition on standardized aptitude and achievement tests. Thus, a promising alternative measure would be one that could predict academic achievement for members of underrepresented groups with an unusually high likelihood of success. Researchers at the Educational Testing Service generated one such approach when they used background variables like family income and parental education to predict SAT scores. Students whose actual SAT scores were well above what was predicted were labeled "Strivers" (described in the August 31, 1999, issue of the *Wall Street Journal*) and hence were designated as prime targets for admission. In that newspaper interview the project's director, Anthony Carnevale, said, "When you look at a Striver who gets a score of 1000, you're looking at someone who really performs at 1200." Harvard emeritus professor Nathan Glazer, in an article on Strivers in the September 27, 1999, *New Republic* indicated that he shares this point of view when he said (p. 28) "It stands to reason that a student from a materially and educationally impoverished environment who does fairly well on the SAT and better than other students who come from a similar environment is probably stronger than the unadjusted score indicates."

This approach was abandoned once it was recognized that these inferences were incorrect, and such a procedure, if implemented, would yield disappointing results[12]—that in fact "when you look at a Striver who gets a score of 1000, you're looking at someone who really performs at 950."* The misunderstanding that Strivers represented has been known for more than a century as regression to the mean,[13] and more recently as Kelley's paradox.[14] When a prediction system is fallible (its predictions are not perfect), those individuals who are predicted to be far from the mean of their group will, in fact, perform closer to the mean. As an example, if we predicted a child's future height from her parents' height we would find that children of very tall parents will, on average, be tall, but not as tall as their parents. And, similarly, children of very short parents will, on average, be short, but not as short.

An imaginative new approach to balancing minority representation was suggested in a widely cited 2003 paper by Roy Freedle. Black

*This phrasing is not technically correct, but was used to remain parallel to Carnevale's statement. It would be correct to say, "When you look at Strivers who score 1000 you are looking at a group of individuals who, on average, perform at the 950 level."

examinees score about a standard deviation lower, on average, than (non-Hispanic) White examinees on the Scholastic Assessment Test (SAT). Freedle noticed that if you matched examinees on total SAT score and looked at their performance on easy and hard items separately, a remarkable result ensued. Specifically, on the easy items the White examinees did slightly better than the Black examinees who had the same overall SAT score. Simultaneously, Freedle discovered that Black examinees did slightly better on the hard items than the matched White examinees. Freedle characterized this as an ethnic and social-class bias and proposed to correct it by using just the hard items to construct a "Revised SAT" score, the R-SAT, which would then decrease the racial differences that are typically observed in SAT scores. Freedle's argument was compelling enough to warrant an extended discussion in the *Atlantic Monthly*.[15] In the next section I examine the logic of this approach; then I describe a simulation of a completely fair test that yields results essentially identical to Freedle's. Last, I describe and demonstrate a slight alteration to Freedle's methodology that does not fall prey to the same flaws.

### 8.2. FLAWS IN FREEDLE'S METHOD

There are two parts to Freedle's methodology that are contradictory. The first is his method of stratification on total SAT score, and the second is his drawing of inferences from a division of items into two parts, easy and hard. Bear in mind that if two people have the same SAT score, and one of them did better on the easy items, the other had to do better on the hard ones. If not, they would not have had the same score. This methodology is almost certain to lead us astray. Let us consider a parallel experiment in which we divide people up into pairs, in which each member of each pair is the same height. We then measure the leg length and body length of each person in a pair. We find that if one person has longer legs the other has a longer trunk. How else could they be the same height? If, after we observed this in all pairs, we were to conclude that there is a negative relationship between leg length and trunk length, we would be very wrong. In general, tall people have long legs and big bodies, and short people have short legs and small bodies. But this method of stratifying, and looking within strata, has hidden this from us. Thus Freedle's observation that Black examinees do better on hard items than matched White examinees is not a discovery so much as a tautology that arises from his method of matching.

To understand the pattern of results Freedle observed, it is instructive to consider the nature of the responding process. Answering a particular test question correctly does not depend solely on examinee ability. For multiple choice tests, such as the SAT, if examinees don't know the correct response, they still have a chance of answering the question correctly. Put simply: when we know the answer, we give it; when we don't, we guess. Thus, an examinee's chance of answering an item correctly is more influenced by luck on harder questions than it is on easy questions. Luck in guessing randomly is almost surely distributed equally among all demographic groups, and thus one would not expect an advantage for either Blacks or Whites.

If, however, we stratify examinees based on their total scores, we mix together four different kinds of examinees: (1) those who have relatively high ability and were also lucky when it came to guessing answers to the harder questions, (2) those who have a relatively high ability and were not as lucky, (3) those who have a relatively low ability but were lucky, and (4) those who have a relatively low ability and were not lucky. The examinees in category (1) will get placed into the higher strata and perform similarly well across all items. Likewise, the examinees in category (4) will get placed into the lower strata and perform similarly poorly across all items. However, by means of this stratification method, the examinees from categories (2) and (3) will be placed inadvertently into the same stratum, despite the fact that their abilities actually differ. By virtue of their relatively high ability, the examinees in category (2) will answer easy questions correctly. But, because they were not lucky, they will not have answered many hard questions correctly. The examinees in category (3), however, respond randomly to more questions, and therefore answer some easy questions correctly and some hard questions correctly. Two intertwined statistical artifacts thus appear: examinees in category (2) will consistently answer more easy items correctly, and examinees in category (3) will consistently answer more hard items correctly.

Thus, when we compare the performance of examinees within these strata for easy and hard items separately, the consistent advantage for category (2) examinees on easy items and the corresponding advantage for category (3) examinees on hard items is not surprising. Since White examinees score higher, on average, than Black examinees, it is reasonable to expect more White examinees in category (2) and more Black examinees in category (3). Therefore, according to this analysis at least, the results that Freedle obtained are to be expected. But this predicts only the direction of results, not the size. With a test as reliable as the SAT, we would not expect a large luck effect. Hence what we should observe are subtle differences.

### 8.3. A SIMULATION

To examine the probable size of the Freedle effect we did a simulation that matched, as closely as we could manage, the character of the data set that Freedle used. First we "built" a fair test. It had eighty-five "items" whose difficulty varied over the same range as a typical SAT.* These items were all equally discriminating, and, should any examinees choose to guess, they all had the same likelihood of yielding a correct answer. Next we constructed two groups of "simulees." One group, representing White examinees, was 100,000 strong and their ability was normally distributed around the center of the ability range. The second group, representing Black examinees, contained 20,000 simulees and had the same amount of variance as the first group, but was centered a standard deviation lower. This presents an idealized match to the sort of data Freedle described.

We ran the simulation, collected simulees into score categories, and examined the proportion correct on the forty easiest items for each score category for both groups of simulees. This analysis was then

*For those more technically inclined, the simulation modeled the items using the three-parameter logistic (3-PL) function. All the item discriminations were equal ($a = 1$); the item difficulties were drawn from a normal distribution centered at the same place as the population ability distribution [$b \sim N(0,1)$]. The likelihood of guessing correctly was set at 20% for all items ($c = 0.2$). The Black and White examinees were drawn from normal distributions with the same variance but centered a standard deviation apart [$\theta_{\text{White}} \sim N(0,1)$, ($\theta_{\text{Black}} \sim N(-1,1)$]. We recognize that this simulation is an idealized version of a true SAT. The 3-PL function is a good, but by no means perfect, representation of the responding pattern; slopes are all near $a = 1$, but there is some variation; the lower asymptotes typically vary around a mean of about $c = 0.2$. In addition, the difference in the means of the score distributions is typically a little larger than one standard deviation. But our simulation is meant as a plausible model, not an exact replica. Experience has shown that its results are representative of findings from any of a variety of different SAT data sets.

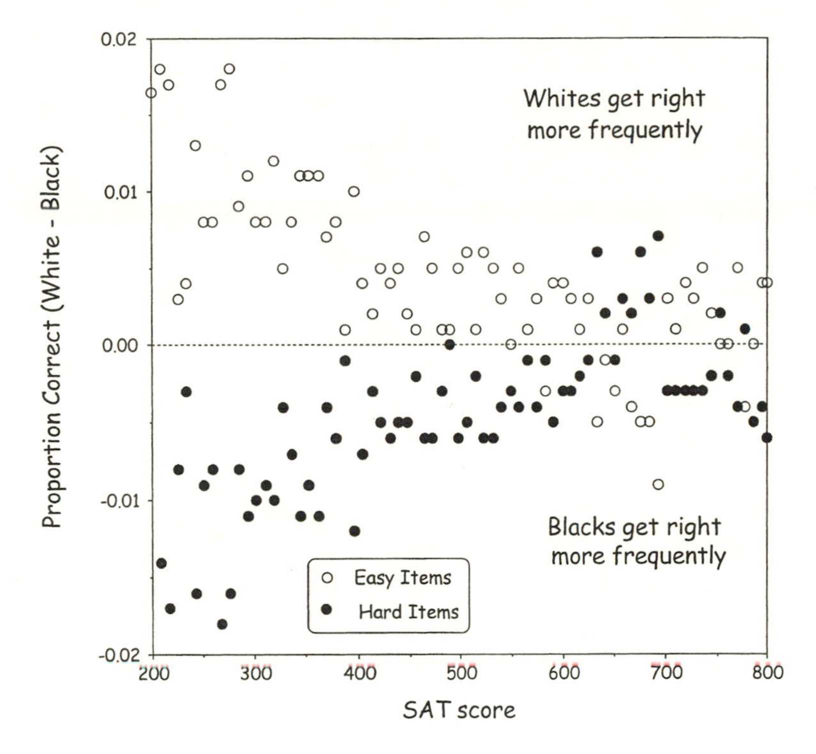


**Figure 8.1.** The results of the simulation for 120,000 simulees. The results for hard and easy items show the expected symmetry and match Freedle's empirical results. The simulation matches Freedle's results both qualitatively and quantitatively.

repeated for the forty hardest items. We then subtracted the proportion correct for the group representing Black examinees from the corresponding proportion for the group representing White examinees. The results are shown in figure 8.1. We see the expected symmetry around the zero line for hard and easy items, indicating that the advantage Whites have on easy items is offset by the corresponding advantage of Blacks on hard items. These results are generated solely from the variability of the responses and the method of conditioning. There is no bias of any kind built into the simulation.

## 8.4. CONCLUDING OBSERVATIONS

We have seen that a simulation of a completely fair test yields essentially the same results as those obtained by Freedle. This does not prove that the kind of effect he discusses does not exist here as well. Our point is that the size of such an effect must be estimated after removing the effect due to a statistical artifact. Since our results match Freedle's in direction and are very close in amount, there is little room left to provide evidence for the sorts of conclusions that Freedle draws.

There is, however, a way to test Freedle's hypothesis with the same data. We note from data included in Freedle's paper that, for both groups, as scores increase so too do scores on the hard items. We can use this fact to repeat Freedle's study without the tautological aspect of his conditioning, by stratifying on a score based solely on the easy items and then determining the proportion of hard items answered correctly. Previously, Black examinees with lower scores for the easy items were in relatively high-score categories because of their higher scores on the hard items. This effect is ameliorated by conditioning only on easy-item performance, thus "leveling the playing field" for those examinees helped or hurt by guessing. Keeping in mind the fact that higher-scoring individuals of both races do better on both easy and hard items, if Freedle's "Black advantage on hard items" hypothesis is correct, we should see the relative advantage of Blacks over Whites on the hard items increase.

When we implemented this in our simulation we found exactly the opposite; when we matched on easy-item score (and release examinees from the requirement that their easy-item and hard-item scores must compensate for one another), White simulees at all score levels outperformed Black simulees for the hard items (see figure 8.2). Note that the group differences increase as score levels increase, suggesting that if we implement Freedle's idea we are not likely to have the result he envisions. Of course this is only a simulation, and hence it tells us only what would happen under the conditions of a completely unbiased test. To know what happens in reality we would need to repeat this analysis on operational SAT data. In view of the comparability between Freedle's results and those from our simulation in figure 8.1, we are not sanguine that subsequent reanalysis will be very different from what we predict here. In fact, using actual SAT data drawn from tests administered in the fall, winter, and spring of 1994–95, when we matched on easy items and looked at Black/White mean differences on the hard half test, the results were as predicted by the simulation. Below 290 on the easy verbal test there were no systematic Black/White differences on the hard test;* above this point, White

*Examinees in both groups appear to be guessing at random at this score level.

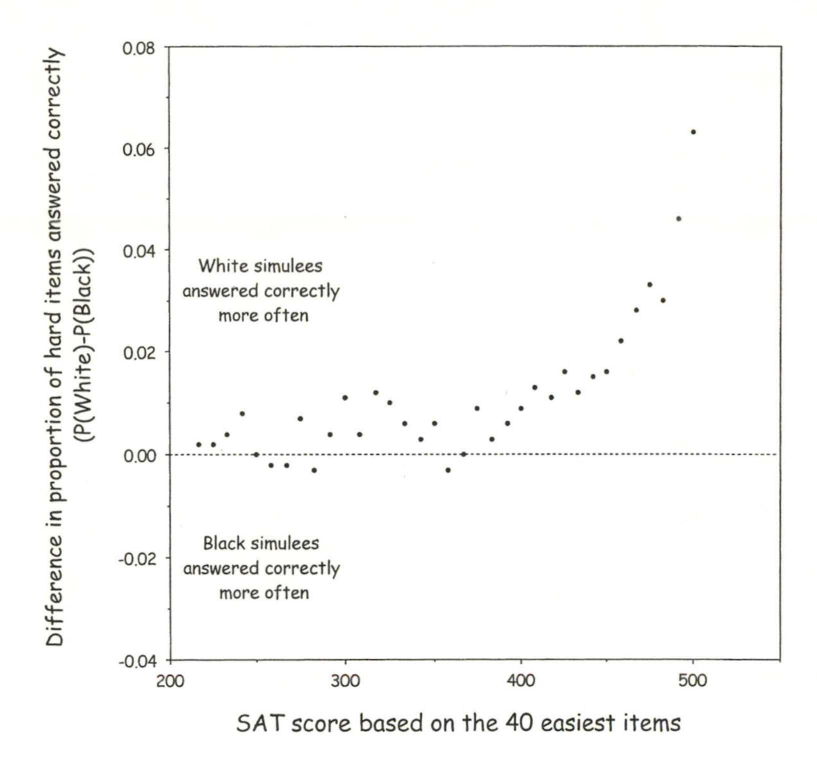


**Figure 8.2.** White simulees answer more hard items correctly than Black simulees who match them on their "easy-item SAT score." When we stratify on a SAT score based solely on the easy items, we find that White simulees outperform Black simulees on the hard items.

students consistently scored at least ten SAT points higher on the hard test. The structure of difference on the math exam was slightly different, in that higher scores on the easy-item math test yielded a greater racial difference on the hard items—an increase of one hundred SAT points on the easy item score yielded an increase of five SAT points in the difference between the races. This difference, contrary to Freedle, was always in favor of White examinees (see figures 8.3 and 8.4).

I have not commented on any of Freedle's suggestions for amending the SAT, for they all lean on his empirical findings for support. Since his empirical findings are completely explained as a statistical artifact on an absolutely fair test, it does not make much sense to discuss how those results might be used to reduce group differences in test scores. I would be remiss, and unfair to Freedle, if I didn't mention that his fundamental plan of building a new SAT on hard items would reduce group differences. However, this reduction does not come from making

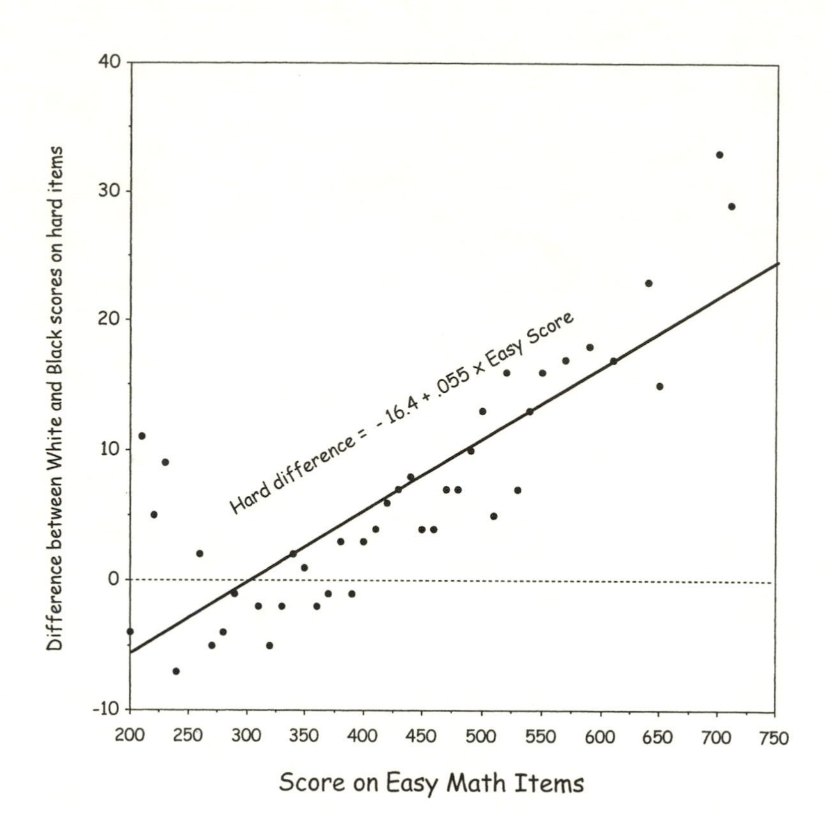


**Figure 8.3.**
The difference between scores obtained by Black and White examinees on hard math items increases by about five points for every one hundred points of difference on easy math items. The actual difference in average scores between White and Black examinees on the math SAT. Sample sizes were 316,791 for White examinees and 37,129 for Black examinees. All differences observed are statistically significant ($p < 0.001$, using Bonferroni adjustment to account for multiple tests).

the test fairer, but instead from making it more error-laden.* Indeed we could eliminate group differences entirely if we based college admission on social security number instead of SAT score. This would be completely random and hence show neither any group differences

*Although this is outside the principal flow of our argument, we felt an *obiter dictum* on test fairness was worth inserting here. A test's fairness ought not be judged by the size of the intergroup differences—a ruler would not be considered fair if it showed that adults' and children's heights were the same—and so simply showing that group differences are smaller is not evidence that the test is fairer. To do this requires that we show that the size of the differences portrayed is a more accurate reflection of the true differences between the groups. And so, before accusing the SAT of cultural bias because of the observed intergroup differences, one must examine validity data. The SAT is meant as a tool to aid college admission officers in predicting prospective students' performance in their freshman year. As such a tool the SAT, a three-hour test, is about as accurate a predictor as four years of high-school grades. If it was found to underpredict the subsequent performance of minority students in college there would be evidence of bias. In fact, it tends to substantially *overpredict* minority performance (e.g., Bowen and Bok, 1998, figure 3.10; Ramist, Lewis, and McCamley-Jenkins, 1994). One cause of this overprediction has been described as Kelley's paradox (Wainer, 2000b), but will not be elaborated upon here.

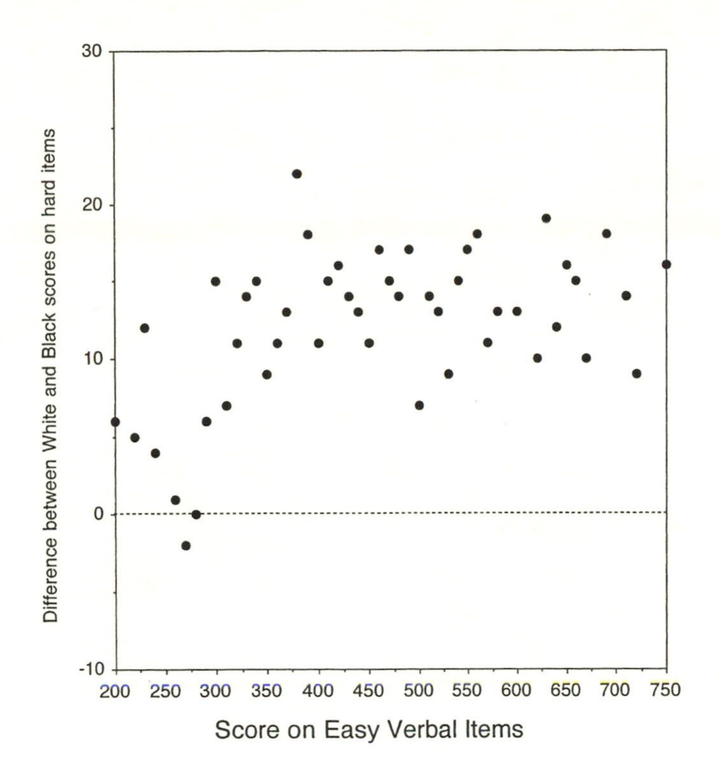


**Figure 8.4.**
The actual difference in average scores between White and Black examinees on the verbal SAT. There is about a fifteen-point difference between scores obtained by Black and White examinees on hard verbal items for examinees with scores on the easy items over three hundred. Sample sizes were the same as in the math portion. All differences observed are statistically significant ($p < 0.001$, using Bonferroni adjustment to account for multiple tests).

nor any validity; thus, while satisfying part of the new face of fairness, it violates the key restriction.

> The new face of fairness encourages policy makers and measurement professionals to shape decisions that will, *within the bounds of validity,* reduce group differences in scores. (Cole and Zieky [2001], p. 369)

Let me end this somewhat technical discussion with a clear and practical conclusion. Roy Freedle's plan to reduce group differences on the SAT while maintaining the test's usefulness as a predictor of future college performance by increasing the difficulty of the exam is silly in concept and it won't work in practice.

# 9

## Insignificant Is Not Zero: Musing on the College Board's Understanding of Uncertainty

### 9.1. INTRODUCTION

On October 8, 2005, NCS Pearson, Inc., under contract to the College Entrance Examination Board, scored an administration of the SAT Reasoning test. Subsequently it was discovered that there was a scanning error that had affected 5,024 examinees' scores. After rescoring it was revealed that 4,411 test scores were too low and 613 were too high. The exams that were underscored were revised upward and the revised scores were reported to the designated colleges and universities. The College Board decided that "it would be unfair to re-report the scores of the 613 test takers" whose scores were improperly too high and hence did not correct them. A motion for a preliminary injunction to force the rescoring of these students' tests was then filed in the United States District Court for the District of Minnesota (Civil Action no. 06-1481 PAM/JSM).*

The College Board's curious decision to leave erroneous scores uncorrected was made on the basis of a mixture of reasons: some practical

*The affidavits on which this essay is based are all publicly available documents. They can be obtained directly from the U.S. District Court of Minnesota and they can also be accessed from http://pacer.uspci.uscourts.gov/ which requires a modest fee. An alternative route is through http://www.mnd.uscourts.gov/cmecf/index.htm which also requires a small fee.

and some statistical. In this chapter I will examine some of their statistical arguments.

The cause of the scoring errors was described as a mixture of two factors, excess humidity that apparently caused some answer sheets to expand and hence not be properly scanned, and too light shading by some examinees of the answer bubbles. Apparently the underscoring occurred when correct answers were missed by the scanner. Overscoring occurred when incorrect answers were not sensed and so were treated as omitted items. Because the SAT is "formula scored," one-fourth of the wrong answers are deducted from the number of correct answers (to adjust for guessing). Thus missing incorrectly marked answers yielded a small boost in score.

The College Board reported that "fewer than 5 percent of the 4,411 (too low) scores were understated by more than 100 points. A total of only ten scores were under-reported by more than 300 points." They also reported that "550 of 613 test takers had scores that were overstated by only 10 points; an additional 58 had scores that were overstated by only 20 points. Only five test takers, . . . , had score differences greater than 20 points" (three had 30 point gains, one 40, and one 50).

### 9.2. WHY NOT CORRECT THE ERRORS?

What were the statistical arguments used by College Board not to revise the erroneously too high scores?

(1) "None of the 613 test takers had overstated scores that were in excess of the standard error of measurement" or (in excess of)
(2) "the standard error of the difference for the SAT."
(3) "More than one-third of the test takers—215 out of 613—had higher SAT scores on file"

And in addition they had a few other arguments that were not quantitative, specifically:

(4) "More than one-fourth of the test takers—157 out of 613—did not ask the College Board to report their SAT scores."
(5) "Not all test takers were seniors."

It seems sensible to discuss each of these, one at a time, but first let us frame the discussion within the context of the purposes of testing.

Tests have three possible goals: they can be used as measuring instruments, as contests, and as prods. Sometimes tests are asked to serve multiple purposes, but the character of a test required to fulfill one of these goals is often at odds with what would be required for another. However, there is always a primary purpose, and a specific test is usually constructed to accomplish its primary goal as well as possible.

A typical instance of tests being used for measurement is the diagnostic tests that probe an examinee's strengths and weaknesses with the goal of suggesting remediation. For tests of this sort the statistical machinery of standard errors is of obvious importance, for one would want to be sure that the weaknesses noted are likely to be real so that the remediation is not chasing noise.

A test used as a contest does not require such machinery, for the goal is to choose a winner. Consider, as an example, the Olympic 100 meter dash—no one suggests that a difference of one-hundredth of a second in the finish portends a "significant" difference, nor that if the race were to be run again the same result would necessarily occur. The only outcome of interest is "who won." Of key importance here is that, by far, the most important characteristic of the test as contest is that it be fair; we cannot have stopwatches of different speeds on different runners.

### 9.3. DO THESE ARGUMENTS HOLD WATER?

The SAT is, first and foremost, a contest. Winners are admitted, given scholarships, etc., losers are not. The only issue regarding error that is of importance is that you have measured accurately enough to know unambiguously who won. For example, recently, Justin Gatlin was initially deemed to have broken the world record in 100 meters (9.76 seconds vs. the old record held by Asafa Powell of 9.77). Subsequently this decision was changed when it was revealed that the runner's actual time was 9.766 and it was incorrectly rounded down. This decision explicitly confirms that differences of thousandths of seconds were too small to allow the unambiguous choice of a winner. The arguments of the College Board about the insignificance of score differences of ten or twenty or thirty points are directly analogous. In what follows we will provide both statistical arguments about the viability of the College

Board's explanations as well as evidence that such score differences are too large to be ignored.

The College Board's argument that, because the range of inflated test scores is within the standard error of measurement or standard error of difference, the scoring errors are "not material" is specious. The idea of a standard error is to provide some idea of the stability of the score if the test should be taken repeatedly without any change in the examinee's ability. But that is not relevant in a contest. For the SAT what matters is not what your score might have been had you taken it many more times, but rather what your score actually was—indeed, what it was in comparison with the others who took it and are in competition with you for admission, for scholarships, etc.

Moreover, the standard error reported by the College Board is incorrect for the individuals whose tests were mis-scored because the scoring error is in addition to the usual standard error. But this sort of error is different. The standard error is thought of as symmetric in that the test taker's observed score is equally likely to be above or below her true score.* The scoring errors were asymmetric, only adding error in the positive direction, thus adding bias as well as error variance. These comments are equally valid for the standard error of the difference, which is based on the standard error.

The College Board's arguments about the standard error of the difference are incorrect on many bases. They are incorrect logically, because the test is used as a contest and so the significance of the difference under the counterfactual situation of repeated testing is irrelevant. They are incorrect technically, because the standard error used to calculate them is incorrectly small and does not reflect the bias in the actual situation. And they are incorrect statistically, in that even if two individuals are not "significantly different" it does not mean that the best estimate of their difference is zero. If two candidates are, for example, twenty points apart, we may not be able to dismiss entirely the possibility that they are of equal ability, but nevertheless our best estimate of their relative ability is that they are about twenty points apart.

*"True score" is not some Platonic score obtained by looking deep within the soul of the examinee; rather it is psychometric jargon for the average score that an examinee would get if she retook parallel forms of the same exam infinitely many times.

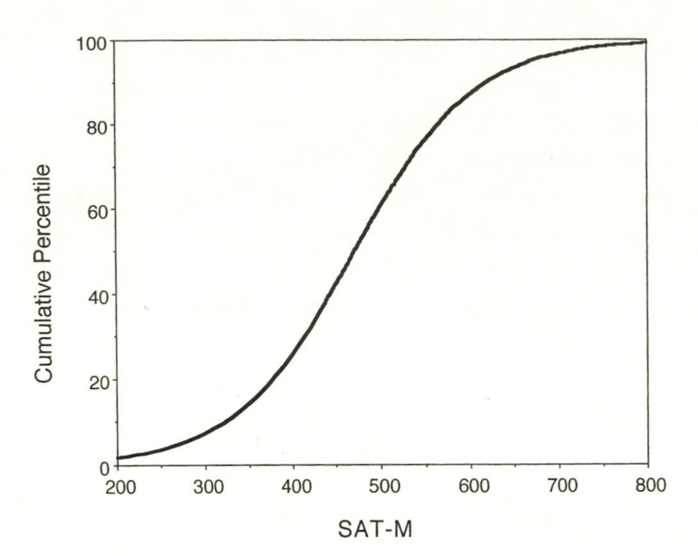


**Figure 9.1.**
Cumulative distribution for the SAT-M 2005 data.

Hence a sensible decision process would choose, *ceteris paribus,* the candidate with the higher score.

### 9.4. HOW BIG IS THE EFFECT?

So far my argument has been on the logical bases of the College Board's decision not to rescore the erroneously high SAT scores. One might decide that, although the scores are in error, the impact the errors would have is so small that the sin of leaving them uncorrected is venial, not mortal. But how small are they? One measure of their effect might be the number of people jumped over by the test taker if given an unearned boost of ten points. We could easily estimate this if we could somehow construct the cumulative distribution of test scores. The College Board publication "2005 College-Bound Seniors: Total Group Profile Report" provides some information that can help us to construct this distribution. Table 6 in that publication yields twelve score categories with the proportion of test takers in each. From that information I constructed the cumulative distribution shown in figure 9.1.

With the cumulative distribution in hand we can interpolate between the given values to obtain the cumulative percentile for any

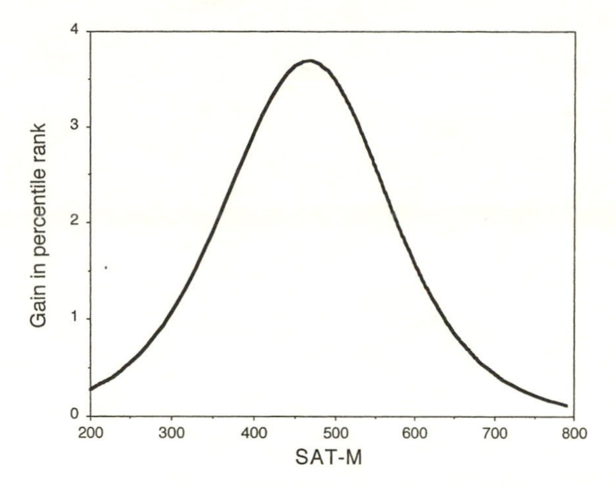


**Figure 9.2.**
Percentile gain for any score if ten points were to have been added to it.

SAT-M score. To look at the effect of a ten-point gain let us examine the change in percentile ranking for a ten-point gain. This is shown in figure 9.2.

The results shown there tell us many things. Two of them are that (i) a ten-point gain in the middle of the distribution yields a bigger increase in someone's percentile rank than would be the case in the tails (to be expected since the people are more densely packed in the middle and so a ten-point jump carries you over more of them in the middle), and (ii) the gain can be as much as 3.5 percentile ranks.

But percentiles do not live near everyone's soul; a more evocative picture is yielded by transforming the scale to people. We know that almost one and a half million students took the SAT-M in 2005 (actually 1,475,623), and so if we multiply the percentiles in figure 9.2 by this number we get the same picture scaled in units of people. This is shown in figure 9.3.

And so we see that very low- or very high-scoring examinees will move ahead of only four or five thousand other test takers through the addition of an extra, unearned, ten points. But someone who scores near the middle of the distribution (where, indeed, most people are) will leapfrog ahead of as many as 54,000 other examinees.

We can use the results shown in figure 9.3 to calculate how many students have been potentially disadvantaged by the College Board's

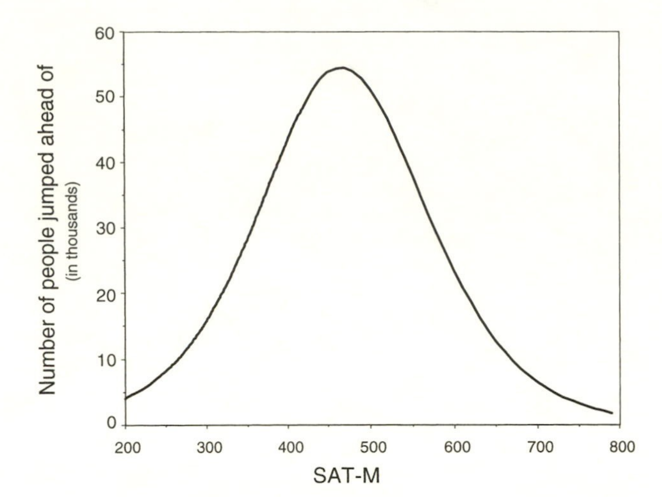


**Figure 9.3.**
A ten-point gain in SAT-M score can mean that you have jumped ahead of as many as 54000 other test takers.

decision not to correct their scoring error. We do this by inserting the 613 overscored individuals into this figure (on the *x*-axis) and adding together all of the associated effects (shown on the *y*-axis). Without knowing those specific 613 scores (even the actual students affected don't know) we cannot calculate the total effect, but it is likely that this will add up to hundreds of thousands. The structure shown here for the SAT-M is paralleled by the cumulative distribution of the SAT-V. One would be hard pressed to call such an effect "entirely irrelevant."[1]

I haven't tried to respond to the last three of the College Board's arguments. The last two are the toughest because it is hard to understand any ethical reason why it would be OK to report erroneous scores to examinees who are not seniors or who have not yet chosen to have their scores reported.

A different mystery emerges when we think about the idea of it being OK to report an erroneous score if a higher score is already on file. Many colleges, when a student's record has multiple SAT scores, only count the highest ones. This is done apparently even when it comes to choosing the highest SAT-V from one administration and the highest SAT-M from another. Such a practice collapses into a single category one student who took the SATs once and scored 1600, with another, who after many tries eventually accumulated an 800 SAT-M at one administration and an 800 SAT-V on another. Obviously, if one had to make a choice as to which student was the more able, it would be the first one.* It is thus puzzling that schools that complain about the SAT's lack of discrimination among the very best students use such a procedure, and stranger still that such procedures are encouraged by the College Board.

### 9.5. CONCLUSION

I have focused here primarily on those of the College Board's statistical arguments that might be illuminated with data. But I believe that it was worth going over what is surely old ground (e.g., the meaning of the standard error of a difference) if for no other reason than that if a largely statistical organization, like the College Board, can have such errors of understanding there are surely many others who also believe that if a difference isn't significant it is zero. In 2005, Stanford statistician David Rogosa, in discussing the same misuse of the standard error argument by the *Orange County Register*, characterized this as "the margin of error folly: If it could be, it is." I wish I could've said it as well.

To emphasize this point let me propose a wager aimed specifically at those who buy the College Board argument. Let us gather a large sample, say 100,000 pairs of students, where each pair go to the same college and have the same major, but one has a 30 point higher (though not statistically significantly different) SAT score. If you truly believe that such a pair are indistinguishable you should be willing to take either one as eventually earning a higher GPA in college. So here is the bet—I'll take the higher-scoring one (for say $10 each). The amount of money I win (or lose) should be an indicator of the truth of the College Board's argument.

If I can get enough people to take this bet my concerns about financing a comfortable retirement will be over.

### 9.6. A LEGAL POSTSCRIPT

Thus far I have presented my own arguments in contrast to those proposed in court documents and other formal publications by the College

*Consider a dice game where the goal was to get the highest total score on the two dice. Suppose the rules allowed someone to toss the dice several times (with a small fee for each toss) until one 6 showed up. Then continue tossing until the second die came up 6. And then claim 12 as your total. How much would a casino charge for each throw (relative to the prize of winning)? Compare the cost of taking the SAT with what is popularly considered the value of admission to a top college.

Board and NCS Pearson. I had hoped that logic and reason would sway your opinion in my direction. Happily I can now add one last bit of independent evidence to support my point of view.

An August 24, 2007, Associated Press article by Brian Bakst reported a "$2.85 million proposed settlement announced Friday (August 24, 2007) by parties in the federal class-action lawsuit. The payout is by the not-for-profit College Board and test scoring company NCS Pearson." The settlement needs ratification by Judge Joan Ericksen during a hearing scheduled for November 29, 2007.

Edna Johnson, a spokeswoman for the College Board, said, "We were eager to put this behind us and focus on the future." NCS spokesman Dave Hakensen said the company declined comment.

# PART IV Mostly Methodological

This section is a bit more technical than the others, explicitly discussing the statistical tools used to help us make decisions under uncertainty; of course they are discussed with a practical purpose, but the tool is given more of the spotlight. Since the time of Benjamin Disraeli, statistics has always been looked upon with distrust. In chapter 17 of his marvelous *Life on the Mississippi*, Mark Twain poked fun at linear extrapolation when he described how the river would sometimes cut across meanders and hence shorten itself.

> Since my own day on the Mississippi, cut-offs have been made at Hurricane Island; at island 100; at Napoleon, Arkansas; at Walnut Bend; and at Council Bend. These shortened the river, in the aggregate, sixty-seven miles. In my own time a cut-off was made at American Bend, which shortened the river ten miles or more.
>
> Therefore, the Mississippi between Cairo and New Orleans was twelve hundred and fifteen miles long one hundred and seventy-six years ago. It was eleven hundred and eighty after the cut-off of 1722. It was one thousand and forty after the American Bend cut-off. It has lost sixty-seven miles since. Consequently its length is only nine hundred and seventy-three miles at present.
>
> Now, if I wanted to be one of those ponderous scientific people, and "let on" to prove what had occurred in the remote

past by what had occurred in a given time in the recent past, or what will occur in the far future by what has occurred in late years, what an opportunity is here! Geology never had such a chance, nor such exact data to argue from! Nor "development of species," either! Glacial epochs are great things, but they are vague—vague. Please observe:—

In the space of one hundred and seventy-six years the Lower Mississippi has shortened itself two hundred and forty-two miles. That is an average of a trifle over one mile and a third per year. Therefore, any calm person, who is not blind or idiotic, can see that in the Old Oolitic Silurian Period, just a million years ago next November, the Lower Mississippi River was upwards of one million three hundred thousand miles long, and stuck out over the Gulf of Mexico like a fishing-rod. And by the same token any person can see that seven hundred and forty-two years from now the Lower Mississippi will be only a mile and three-quarters long, and Cairo and New Orleans will have joined their streets together, and be plodding comfortably along under a single mayor and a mutual board of aldermen. There is something fascinating about science. One gets such wholesale returns of conjecture out of such a trifling investment of fact.

Linear extrapolation can indeed bring us to some very silly conclusions, but over short distances it is often a fine approximation. But how long is short? In chapter 10 I try to illustrate how to begin to answer this question by looking at the unexpectedly consistent improvements in the world record for men running a mile that have occurred over the course of the twentieth century. I then speculate whether it should have been predictable, and what if anything it means about future improvements in the twenty-first century.

Many sciences, like physics, are so dominated by the specialized language of mathematics that they live at an altitude above the reach of ordinary citizens. If we understand anything at all about such sciences, it is through metaphor and approximation. But the science of uncertainty is closer to where we all live and so not only can its contributions to life be understood broadly but also we can all contribute to its advancement. In chapter 11 I look at statistical graphics in the popular media and show how such graphics, maps to help us understand our uncertain world, have provided models for both good practice and ill. Chapter 12 is a case study demonstrating how a mixture of statistical tools, statistical thinking, and various graphic forms combine to provide us with a step-by-step guided pathway of discovery. The last two chapters are perhaps the most narrowly focused of all, looking first at ways to show our uncertainty graphically and next at one way in which powerful computing when combined with our desire for simplicity at all costs can be used to mislead us. Chapter 13 requires some understanding of traditional hypothesis testing as well as the problems of making multiple comparisons. These issues are also underlying some of the discussion of chapter 14, but much of the technicality is implicit, and a careful reader who is ignorant of such technical methods as are embodied in the Bonferroni inequality can still follow the principal themes. I have also included a short tutorial to aid those readers whose memory of the contents of an introductory statistics course may have dimmed with the passage of time (pages 142–147).

# 10

## How Long Is Short?

All curves are well approximated by a straight line for a short part of their length. Even very wiggly functions can be approximated to a tolerable degree of accuracy by a sequence of straight lines.* Of course, if the portions of the function that can be thought of as linear are very short, parsimony may be better served by joining together longer pieces of relatively simple curves. But how can we know for how long a linear approximation is suitable? Obviously, when the entire data series is in hand it is easy, but what about when it is not? What do we do when we wish to extrapolate from what appears to be linear beyond the data? For a very short extrapolation it is usually fine, but how long is short?

In an essay published in the year 2000 Catherine Njue, Sam Palmer, and I showed how we could be misled by linear extrapolation. We fit linear functions to the winning times in the Boston Marathon (see figure 10.1) and sneered at those who would believe the inferences suggested—specifically that women would be running faster than men by the year 2005.

A year later, when I fit these same data using a smooth interpolating function (see figure 10.2), which is much more sensitive to the local behavior of the data, it became clear that both men and women appeared to be nearing asymptotes. The men's asymptote seemed between two hours and two hours and five minutes, whereas the women's

*Indeed the whole notion of spline functions rests on this idea.

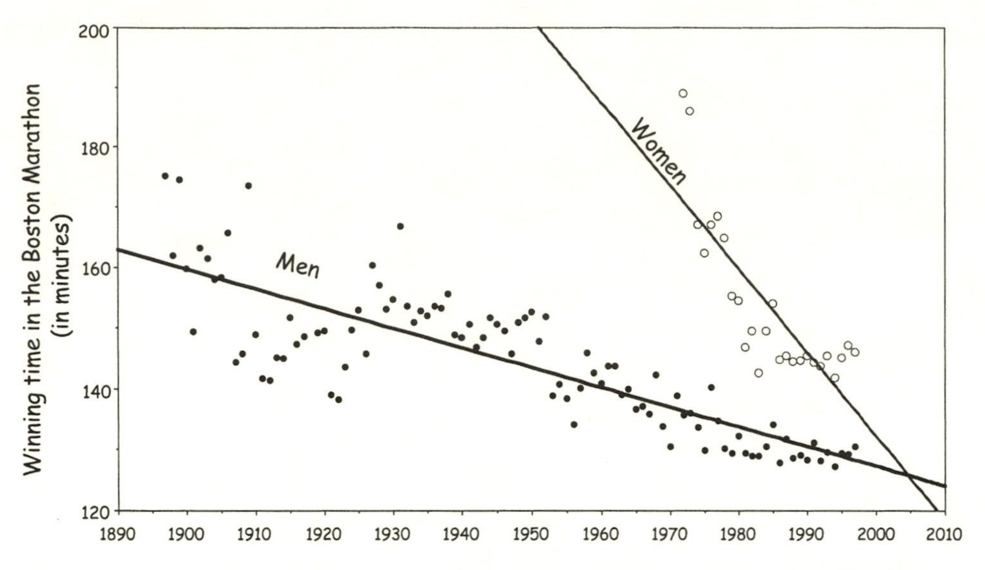


**Figure 10.1.** Linear fit to winning times shows women overtaking men in 2005.

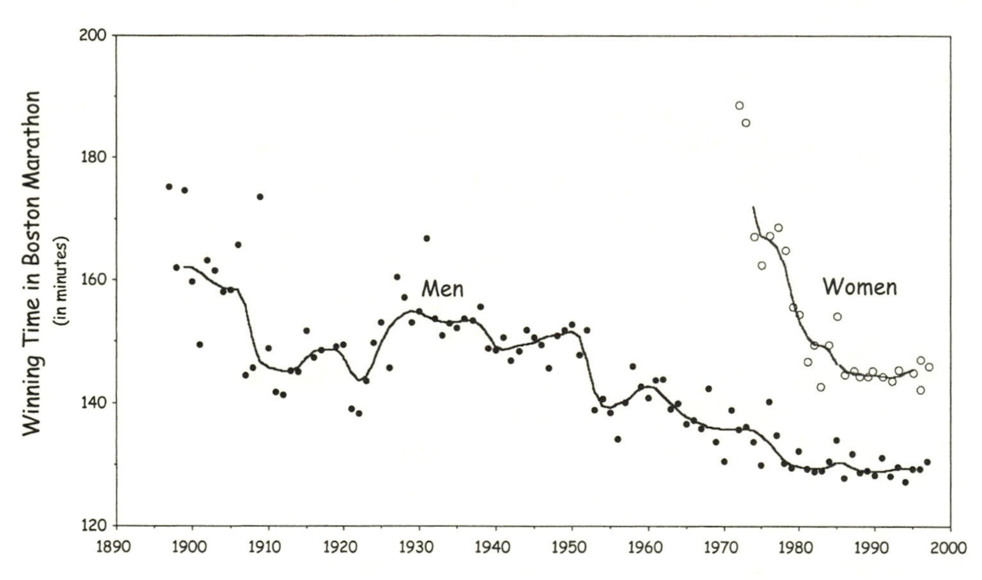


**Figure 10.2.** Smooth fit indicates that both men's and women's times are near an asymptote.

asymptote seemed to be about fifteen minutes slower. I used this example to denigrate the chuckleheads who have the audacity to believe linear extrapolations.

But are such linear extrapolations always a dopey idea? After all, the start of every Taylor expansion* is linear, and there might be circumstances in which we would be surprised by the extent to which linear extrapolation works. Andrew Gelman and Deborah Nolan, in their book on tricks to teach statistics, suggest that we consider the world record times for the mile over the first half of the twentieth century. These are shown in figure 10.3. We see consistent linear improvements of a bit over a third of a second a year, contributed to by such legendary milers as the American Glenn Cunningham, the "Flying Finn" Paavo Nurmi, and other fine Scandinavian runners.

If we used this fitted straight line to predict the path of the world record for the next 50 years (1950–2000) we would find (figure 10.4), remarkably, that the world record for the mile has continued to improve at about the same rate. Indeed, the pace of improvement seems to have even increased a bit. The first forty years of this improvement was the work of the Brits. The English medical student Roger Bannister was the first sub-four-minute miler, but he was followed quickly by the Australian John Landy, who in turn was joined by his countryman Herb Elliott and then the great New Zealand miler Peter Snell. There was a ten-year interlude during which the American Jim Ryun, while barely out of high school, lowered the mark still further, before losing the record to Filbert Bayi, the first African to hold it. Then John Walker, Sebastian Coe, and Steve Ovett reestablished the British Commonwealth's domination of the event. But the decade of the 1990s marked the entrance of North Africans into the competition, as first Noureddine Morceli of Algeria lowered the record, and then the Moroccan Hicham El Guerrouj brought it to its current point of 3:43.13.

Obviously, this linear trend cannot continue forever, but what about for another fifty years? (figure 10.5) It seems unlikely that in

* A Taylor series expansion is a general mathematical method for approximating any differentiable function with a polynomial. It is named after the English mathematician Brook Taylor (1685–1731) who published it in 1714, but its importance was not understood until 1772, when La Grange pointed out that it was an important part of the foundations of differential calculus.

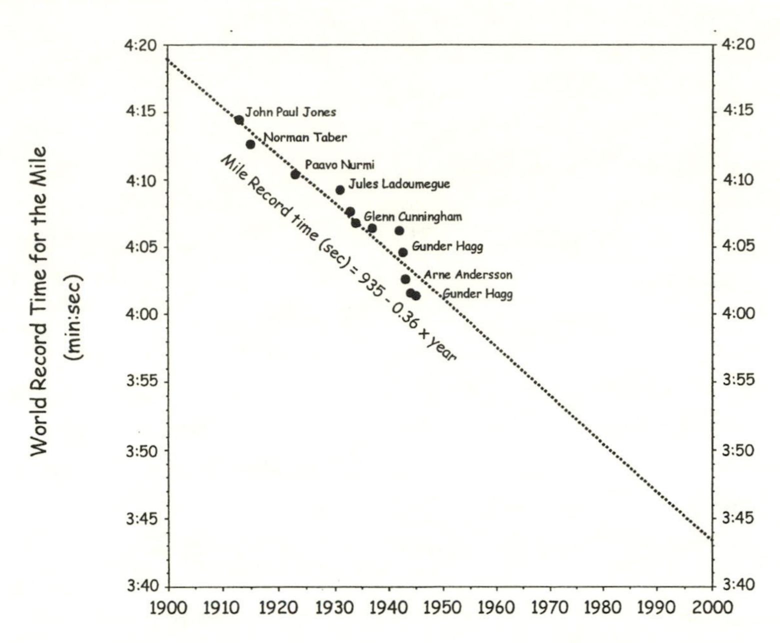


**Figure 10.3.** The world record for the men's mile improved linearly over the first half of the twentieth century.

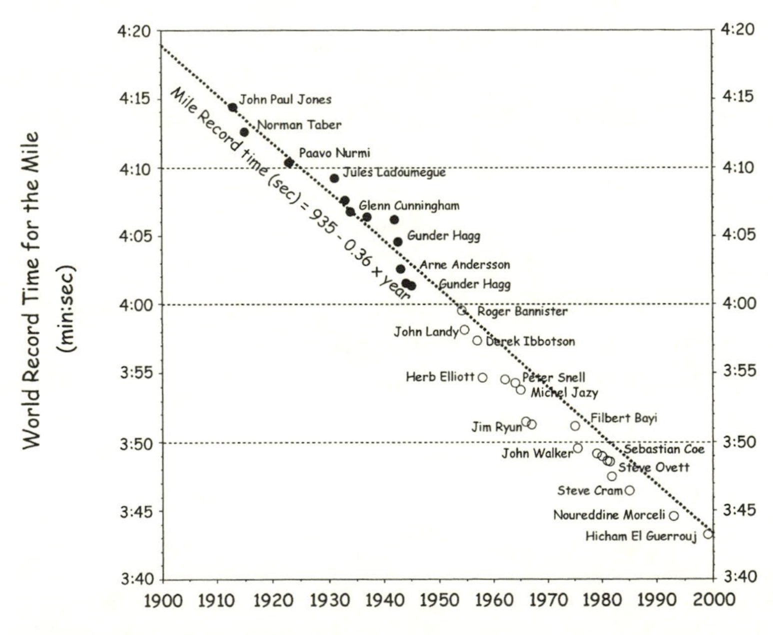


**Figure 10.4.** The linear fit to the data for the first half of the twentieth century provides a remarkably accurate fit for the second half as well.

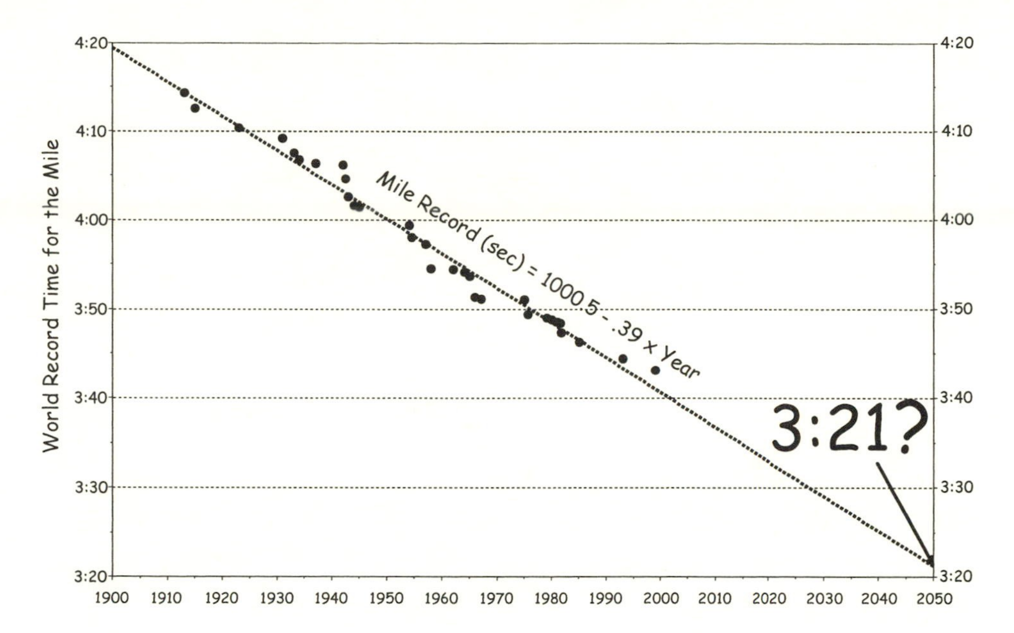


**Figure 10.5.**
For how long can the world record continue to improve at the same rate?

2050 the record for the mile will be as fast as 3:21, but what would we have said in 1950 about a 3:43 mile? Thus the question is not about the linearity of improvement over a short period of time, but rather how long is "short"?

To be able to answer this question we need to understand better the mechanism by which records are set. Scott Berry, in a 2002 article provided one intriguing idea: Specifically, that world records get better because the champion is more highly selected. The best runner out of a billion is bound to be better than the best of a million. His simple model posited that the mean and standard deviation of human ability has remained constant over the past century, but as population increases the best of those competing become further from the mean. This model predicted world records in many areas with uncanny accuracy. Berry's mechanism is clearly operating among milers; as the appeal of the race broadened beyond the domain of western Europeans, the times fell. But why are they falling linearly? Berry's model would suggest that, if participation increased exponentially, times would decrease linearly. There may be minor flaws in the precise specification of Berry's model, but it is hard to argue with the success of its predictions. I think he's on to something.

# 11

## Improving Data Displays

### 11.1. INTRODUCTION

In the transactions between scientists and the media, influence flows in both directions. About twenty-five years ago[1] I wrote an oft-cited article with the ironic title "How to Display Data Badly." In it I chose a dozen or so examples of flawed displays and suggested some paths toward improvement. Two major newspapers, the *New York Times* and the *Washington Post*, were the source of most of my examples Those examples were drawn during a remarkably short period of time. It wasn't hard to find examples of bad graphs.

Happily, in the intervening years those same papers have become increasingly aware of the canons of good practice and have improved their data displays profoundly. Indeed, when one considers both the complexity of the data that are often displayed, as well as the short time intervals permitted for their preparation, the results are often remarkable.

In only a week or two I have picked out a few graphs from the *New York Times* that are especially notable. Over the same time period I noticed graphs in the scientific literature whose data had the same features but were decidedly inferior. At first I thought that it felt more comfortable in the "good old days" when we did it right and the media's results were flawed. But, in fact, the old days were not so good. Graphical practices in scientific journals have not evolved as fast as those of the mass media. It is time scientists learned from their example.

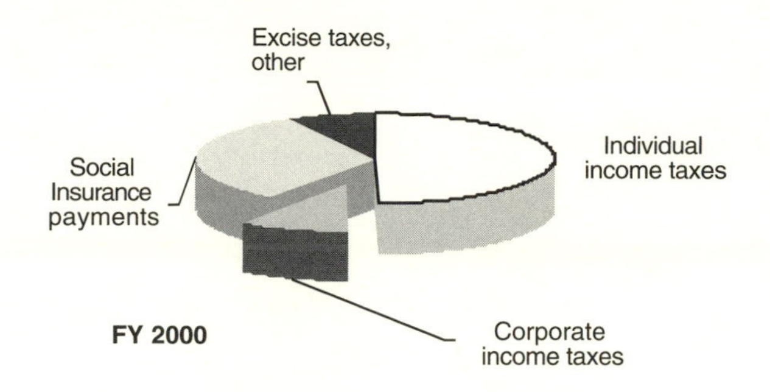


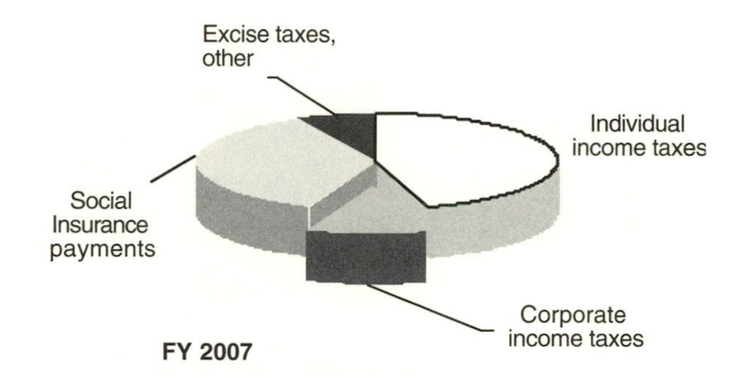


**Figure 11.1.** Federal government receipts by source. A typical three-dimensional pie chart. This one, no worse nor better than most of its ilk, is constructed from U.S. government budget data. http://www.whitehouse.gov/omb/budget/fy2008/pdf/hist.pdf. (From 2007 U.S. Budget Historical Tables.)

## 11.2. EXAMPLE 1: PIES

The U.S. federal government is fond of producing pie charts, and so it came as no surprise to see (figure 11.1) a pie chart of government receipts broken down by their source. Of course, the grapher felt it necessary to "enliven" the presentation with the addition of a specious extra dimension. Obviously the point of presenting the results for both 2000 and 2007 together must have been to allow the viewer to see the changes that have taken place over that time period (roughly the span of the Bush administration). The only feature of change I was able to discern was a shrinkage in the contribution of individual income taxes.

I replotted the data in a format that provides a clearer view (see figure 11.2) and immediately saw that the diminution of taxes was offset by an increase in social insurance payments. In other words, the cost of tax cuts aimed principally at the wealthy was paid for by increasing social security taxes, whose effect ends after the first hundred thousand dollars of earned income.

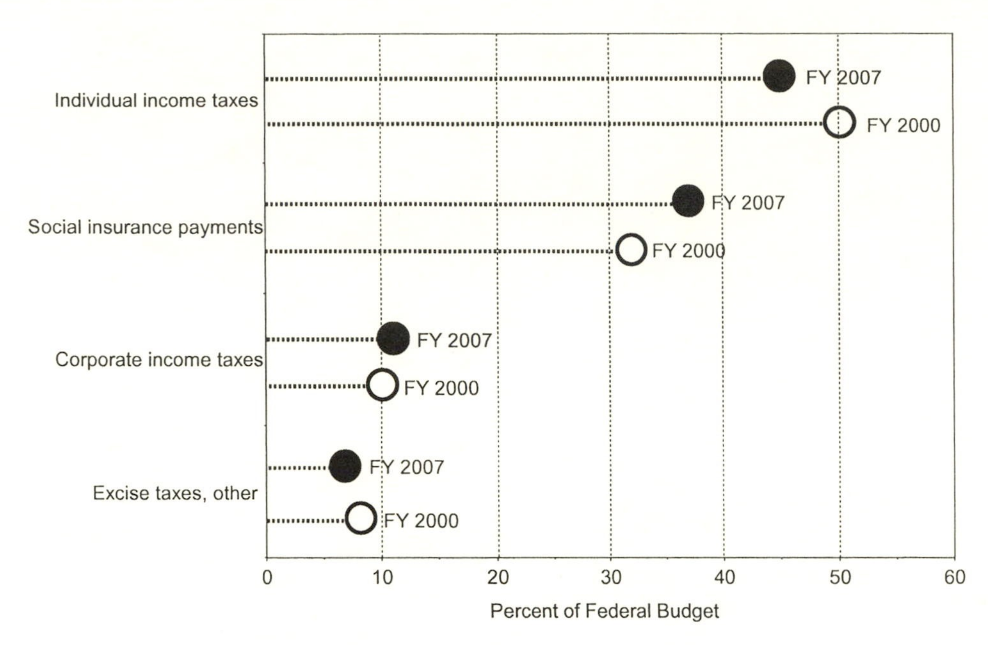


**Figure 11.2.**
Federal government receipts by source. A restructuring of the same data shown in figure 11.1, making clearer what changed in the sources of government receipts between the years 2000 and 2007.

I wasn't surprised that such details were obscured in the original display, for my expectation of data displays constructed for broad consumption was not high. Hence when I was told of a graph in an article in the *New York Times Sunday Magazine* about the topics that American clergy choose for their sermons, I anticipated the worst. My imagination created, floating before my eyes, a pie chart displaying such data (see figure 11.3). My imagination can only stretch so far; hence the pie I saw had but two dimensions and the categories were ordered by size.

What I found (figure 11.4) was a pleasant surprise. The graph (produced by an organization named "Catalogtree") was a variant of a pie chart in which the segments all subtend the same central angle, but their radii are proportional to the amount being displayed. This is a striking improvement on the usual pie because it facilitates comparisons among a set of such displays representing, for example, different years or different places. The results from each segment are always in the same place, whereas with pie charts the locations of all segments vary as the data change. This plot also indicates enough historical consciousness to evoke Florence Nightingale's famous Rose of the Crimean War (see figure 11.5; also chapter 11 in Wainer, 2000).

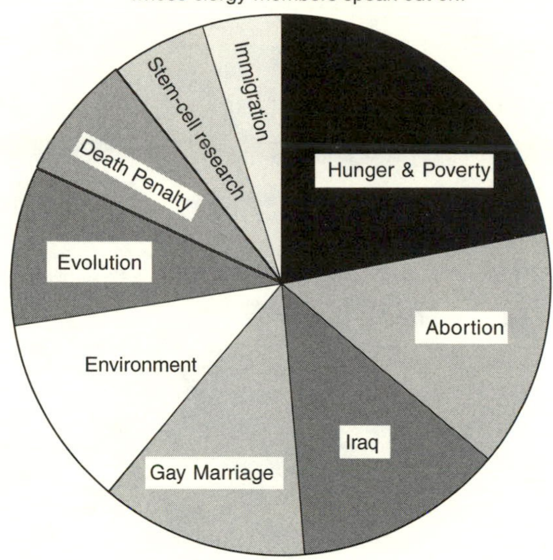


**Figure 11.3.**
Preaching politics. A typical pie chart representation of the relative popularity of various topics among the U.S. clergy.

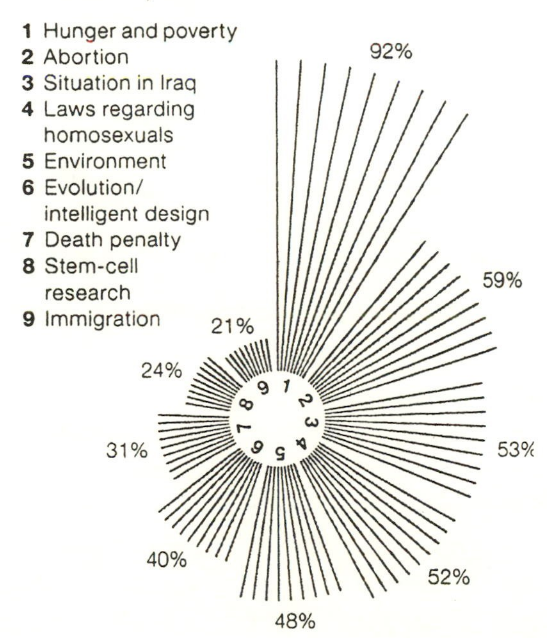


**Figure 11.4.**
A display from the February 18, 2007, *New York Times Sunday Magazine* (p. 11) showing the same data depicted in figure 11.3 as a Nightingale Rose (from Rosen, 2007). Source: August 2006 survey by the Pew Research Center for the People and the Press and the Pew Forum on Religion and Public Life. Numbers based on those who attend religious services at least monthly. Chart by Catalogtree.

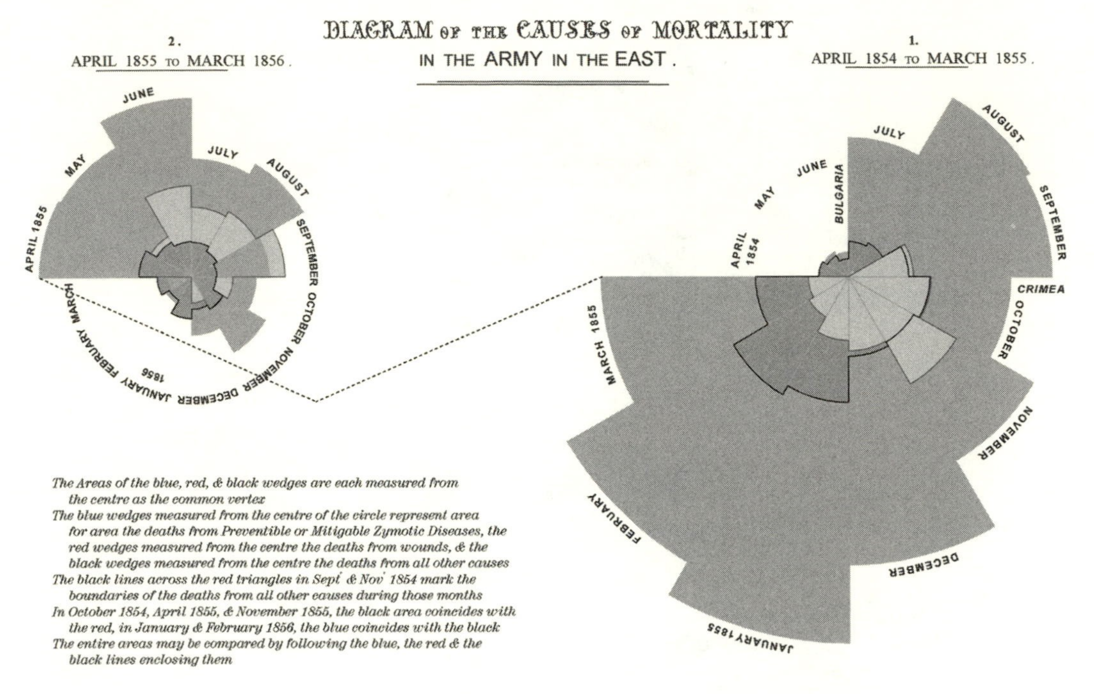


**Figure 11.5.** A redrafting of Florence Nightingale's famous "coxcomb" display (what has since become known as a Nightingale Rose) showing the variation in mortality over the months of the year. (See color version on pages C2–C3.)

Sadly, this elegant display contained one small flaw that distorts our perceptions. The length of the radius of each segment was proportional to the percentage depicted; but we are influenced by the area of the segment, not its radius. Thus the radii need to be proportional to the square root of the percentage for the areas to be perceived correctly. In addition, it is generally preferred if each segment is labeled directly, rather than through a legend. An alternative figuration with these characteristics is shown as figure 11.6.

## 11.3. EXAMPLE 2: LINE LABELS

In 1973 Jacques Bertin, the maître de graphique moderne, explained that when one produces a graph it is best to label each of the elements in the graph directly. He proposed this as the preferred alternative to appending some sort of legend that defines each element. His point was that when the two are connected you can comprehend the graph in a single moment of perception as opposed to having to first look at

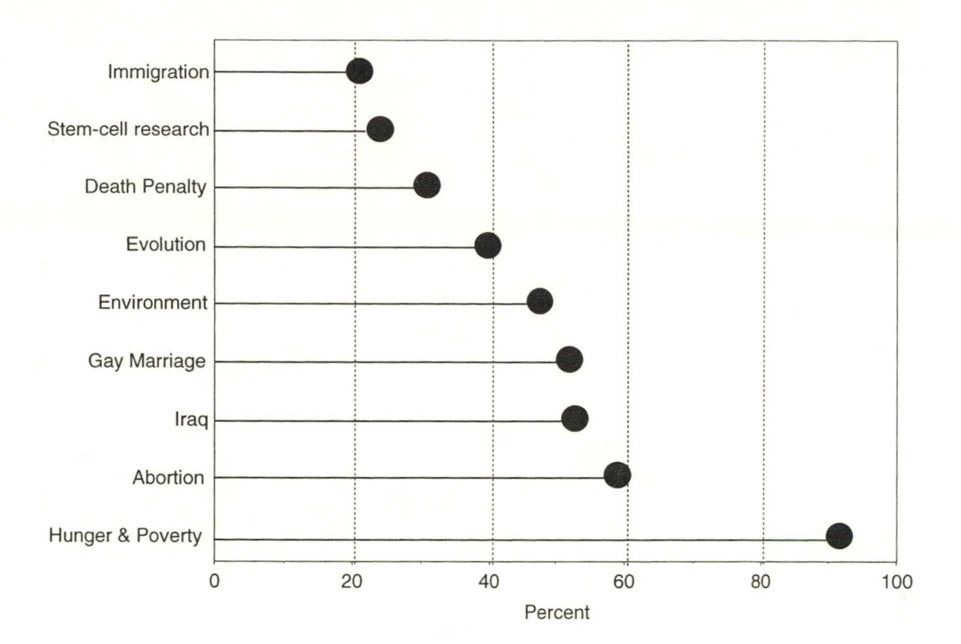


**Figure 11.6.**
The same data previously shown in figures 11.3 and 11.4 recast as a line-and-dot plot.

the lines, then read the legend, and then match the legend to the lines.

Despite the authoritative source, this advice is too rarely followed. For example (see figure 11.7), Mark Reckase (2006), in a simple plot of two lines , chose not to label the lines directly—even though there was plenty of room to do so—and instead chose to put in a legend. And the legend reverses the order of the lines, so the top line in the graph becomes the bottom line in the legend, thus increasing the opportunities for reader error.

Can the practice of labeling be made still worse? figure 11.8, from Pfeffermann & Tiller (2006), shows a valiant effort to do so. Here the legend is hidden in the figure caption, and again its order is unrelated to the order of the lines in the graph. Moreover, the only way to distinguish BMK from UnBMK is to notice a small dot. The only way I could think of that would make the connection between the various graphical elements and their identifiers worse would be to move the latter to an appendix.

How does the *Times* fare on this aspect of effective display? Very well indeed. In figure 11.9 are two plots describing one hundred years

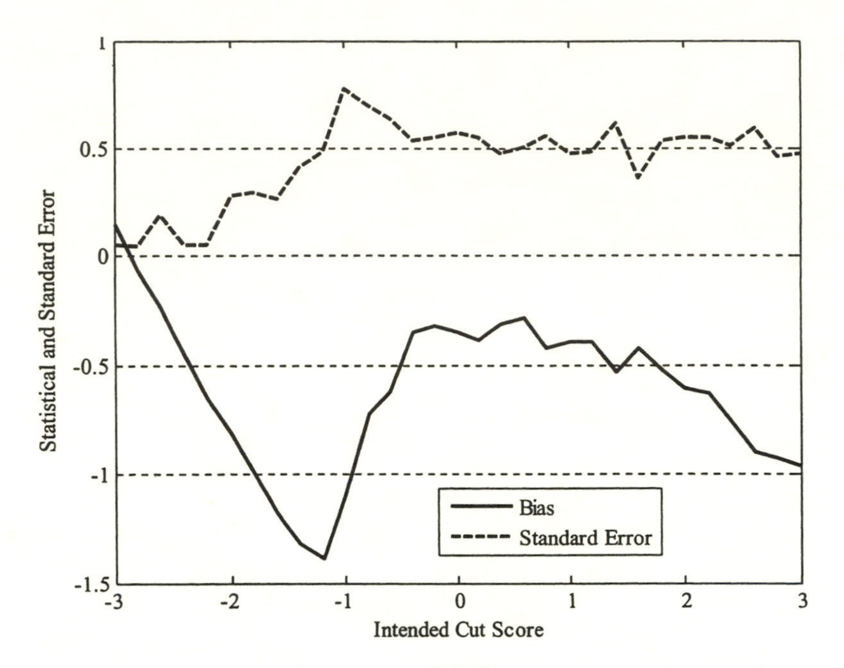


**Figure 11.7.** A graph taken from Reckase (2006) in which instead of the graph lines being identified directly they are identified through a legend—indeed, a legend whose order does not match the data.

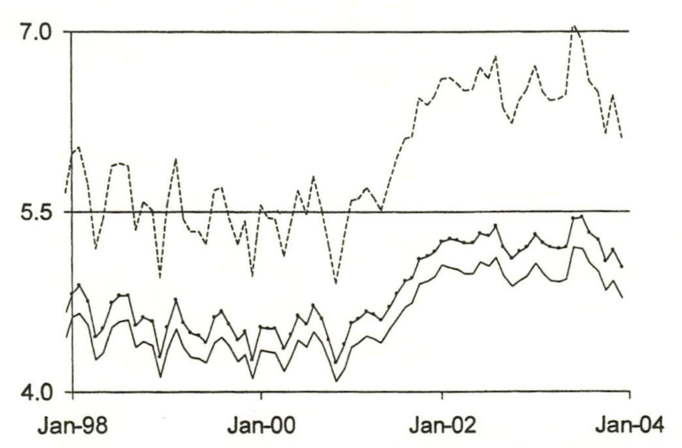


Figure 6. STD of CPS, Benchmarked, and Unbenchmarked Estimates of Total Monthly Unemployment, South Atlantic Division (numbers in 10,000) (------- CPS; —— BMK; —•— UnBMK).

**Figure 11.8.** A graph taken from Pfeffermann and Tiller (2006) in which the three data series are identified in acronym form in the caption. There is plenty of room on the plot for them to be identified on the graph with their names written out in full.

## Shrinking Factory Jobs

The manufacturing industry in New Haven County, Conn., has substantially declined in recent decades, giving way to other industries, like retail and professional services.

**Employment in New Haven County's top three industries**

**Percent of all employment in New Haven County**

200 thousand jobs

60%

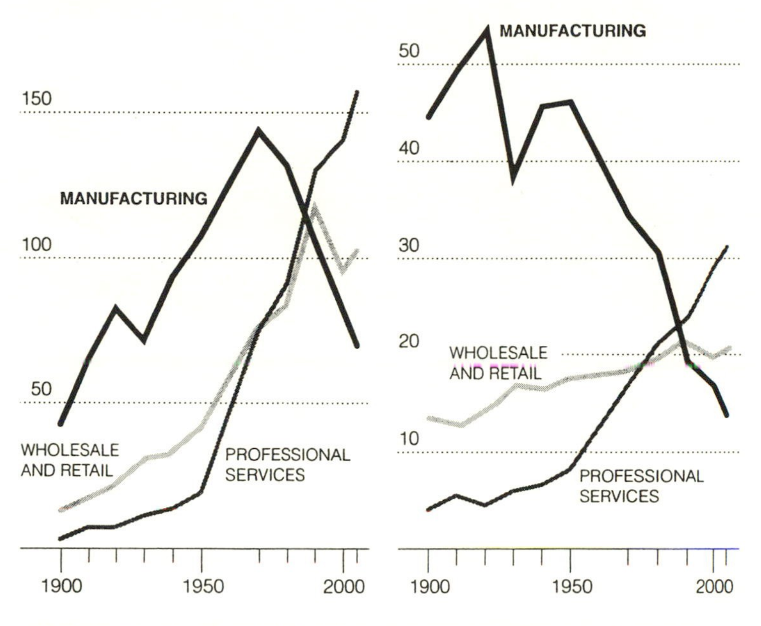


Notes: Data for 1960 is not available. Data for 1980 does not include the towns of Ansonia, Derby and Seymour.

*Sources: Queens College Department of Sociology; Census Bureau*

The New York Times

**Figure 11.9.** A graph taken from the Metro section of the February 18, 2007, *New York Times* showing two panels containing three lines each, in which each line is identified directly.

of employment in New Haven County. In each panel the lines are labeled directly, making the decline of manufacturing jobs clear. In the following week (figure 11.10), another graph appeared showing five time series over three decades. In this plot the long lines and their crossing patterns made it possible for the viewer of it to confuse one line with another. This possibility was ameliorated by labeling both ends of each line. A fine idea worthy of being copied by those of us whose data share the same characteristics.

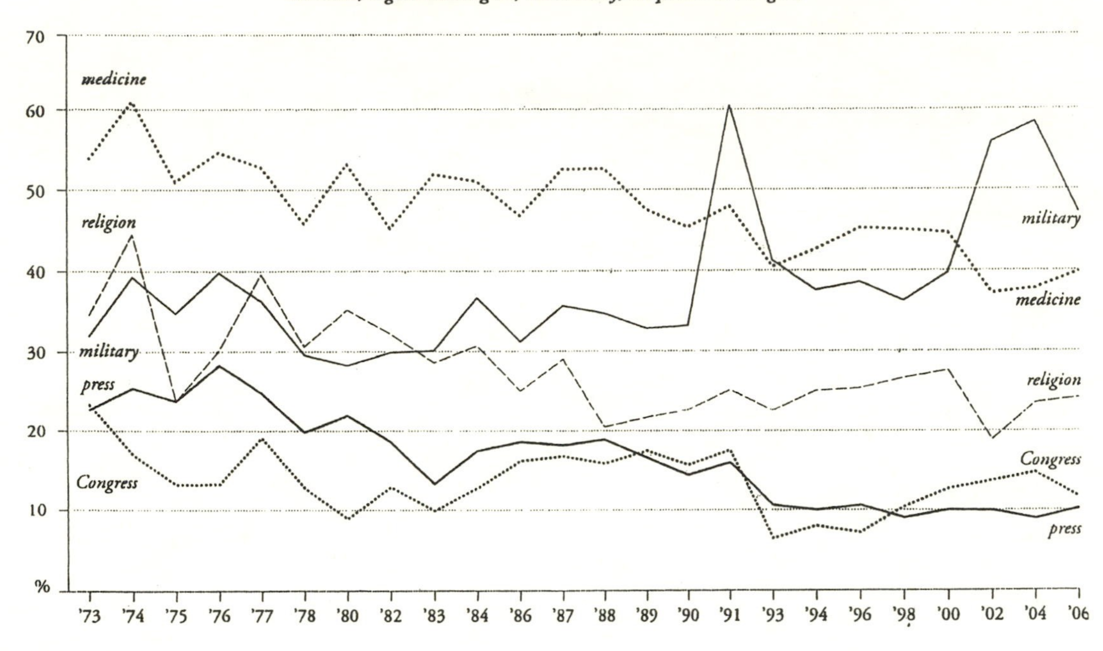


Figure 11.10.
A graph taken from the News of the Week in Review section of the February 25, 2007, *New York Times* (p. 15) showing five long lines each, in which each line is identified directly at both of its extremes, thus making identification easy, even when the lines cross.

### 11.4. EXAMPLE 3: CHANNELING PLAYFAIR

On the business pages of the March 13, 2007, *New York Times* was a graph used to support the principal thesis of an article on how the growth of the Chinese economy has yielded, as a concomitant, an enormous expansion of acquisitions. The accompanying graph, shown here as figure 11.11, has two data series. The first shows the amount of money spent by China on external acquisitions since 1990. The second time series shows the number of such acquisitions. The display format chosen, while reasonably original in its totality, borrows heavily from William Playfair, the eighteenth-century Scottish inventor of many of the graphical forms we use today.[2] First, the idea of including two quite different data series in the same chart is reminiscent of Playfair's (1821) chart comparing the cost of wheat with the salary of a mechanic (figure 11.12, discussed in detail in Friendly and Wainer, 2004). However, in plotting China's expenditures the graphers had to confront the vast increases over the time period shown; a linear scale

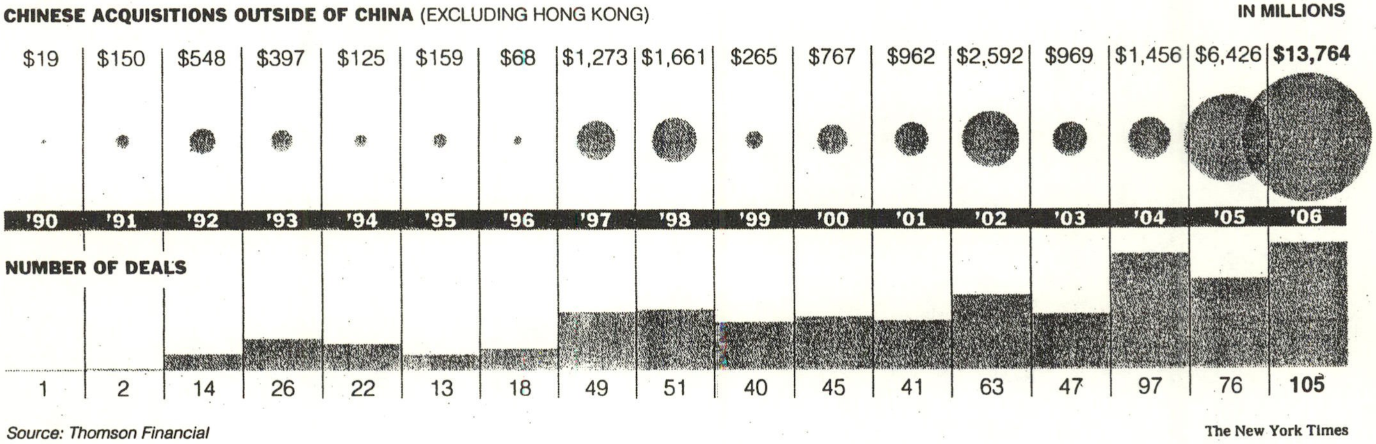

**CHINESE ACQUISITIONS OUTSIDE OF CHINA** (EXCLUDING HONG KONG)

**IN MILLIONS**

| '90 | '91 | '92 | '93 | '94 | '95 | '96 | '97 | '98 | '99 | '00 | '01 | '02 | '03 | '04 | '05 | '06 |
|---|---|---|---|---|---|---|---|---|---|---|---|---|---|---|---|---|
| $19 | $150 | $548 | $397 | $125 | $159 | $68 | $1,273 | $1,661 | $265 | $767 | $962 | $2,592 | $969 | $1,456 | $6,426 | **$13,764** |

**NUMBER OF DEALS**

| '90 | '91 | '92 | '93 | '94 | '95 | '96 | '97 | '98 | '99 | '00 | '01 | '02 | '03 | '04 | '05 | '06 |
|---|---|---|---|---|---|---|---|---|---|---|---|---|---|---|---|---|
| 1 | 2 | 14 | 26 | 22 | 13 | 18 | 49 | 51 | 40 | 45 | 41 | 63 | 47 | 97 | 76 | **105** |

*Source: Thomson Financial*

The New York Times

**Figure 11.11.** A graph taken from the Business section of the March 13, 2007, *New York Times* (p. C1) showing two data series. One is of a set of counts and is represented by bars; the second is money, represented by the areas of circles.

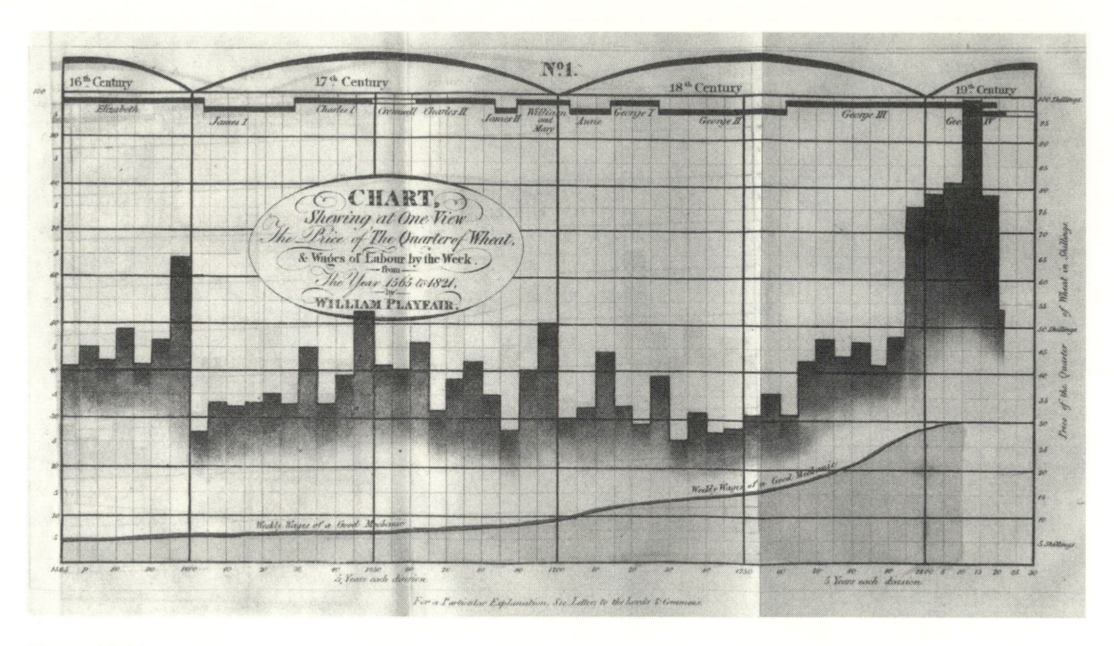


**Figure 11.12.**
A graph taken from Playfair (1821). It contains two data series that are meant to be compared. The first is a line that represents the "weekly wages of a good mechanic" and the second is a set of bars that represent the "price of a quarter of wheat." (See original color version on page C1.)

would have obscured the changes early on. The solution they chose was also borrowed from Playfair's plot of Hindoostan in his 1801 *Statistical Breviary* (figure 11.13). Playfair showed the areas of various parts of Hindoostan as circles. The areas of the circles are proportional to the areas of the segments of the country, but the radii are proportional to the square root of the areas. Thus, by lining up the circles on a common line, we can see the differences of the heights of the circles which is, in effect, a square root transformation of the areas. This visual transformation helps to place diverse data points onto a more reasonable scale.

The *Times*' plot of China's increasing acquisitiveness has two things going for it. It contains thirty-four data points, which by mass-media standards, is data rich, showing vividly the concomitant increases in both data series over a seventeen-year period. This is a long way from Playfair's penchant for showing a century or more, but in the modern world, where changes occur at a less leisurely pace than in the eighteenth-century, seventeen years is often enough. And second, by using Playfair's circle representation it allows the visibility of expenditures over a wide scale.

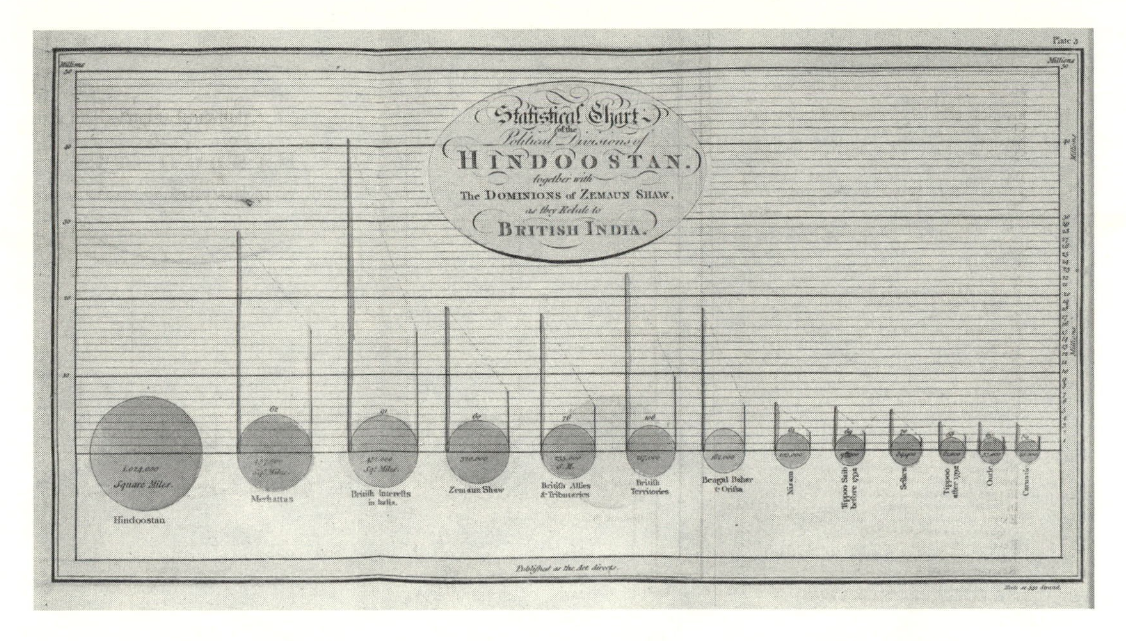


**Figure 11.13.**
A graph taken from Playfair (1801). It contains three data series. The area of each circle is proportional to the area of the geographic location indicated. The vertical line to the left of each circle expresses the number of inhabitants, in millions. The vertical line to the right, represents the revenue generated in that region in millions of pounds sterling. (See original color version on page C4.)

Are there other alternatives that might perform better? Perhaps. In figure 11.14 is a two-paneled display in which each panel carries one of the data series. Panel 11.14a is a straightforward scatter plot showing the linear increases in the number of acquisitions that China has made over the past seventeen years. The slope of the fitted line tells us that over those seventeen years China has, on average, increased its acquisitions by 5.5/year. This crucial detail is missing from the sequence of bars but is obvious from the fitted regression line in the scatter plot. Panel b of figure 11.14 shows the increase in money spent on acquisitions over those same seventeen years. The plot is on a log scale and its overall trend is well described by a straight line. That line has a slope of 0.12 in the log scale, which translates to an increase of about 51% per year. Thus the trend established over these seventeen years shows that China has both increased the number of assets acquired each year and also acquired increasingly expensive assets.

The key advantage of using paired scatter plots with linearizing transformations and fitted straight lines is that they provide a quantitative measure of how China's acquisitiveness has changed. This distinguishes the figure from the *Times* plot, which, although it contained

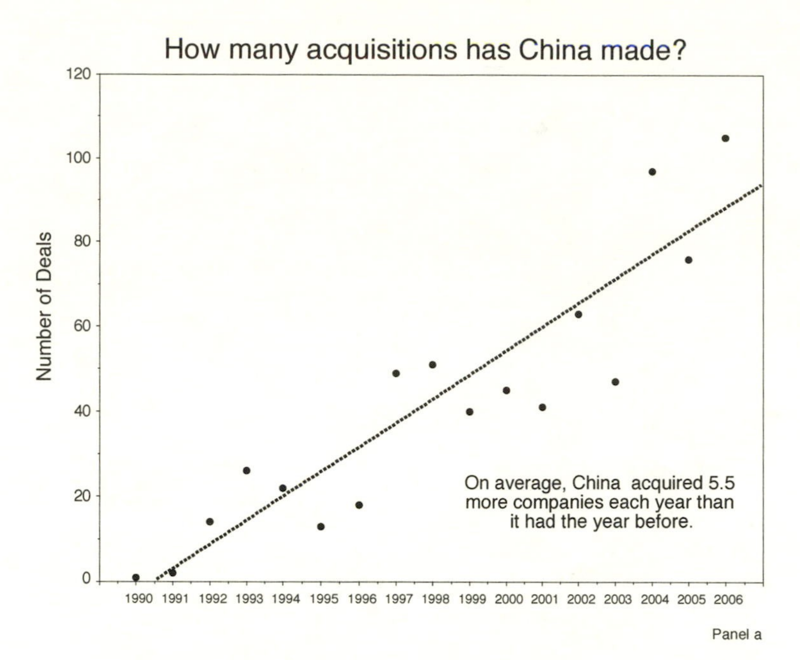


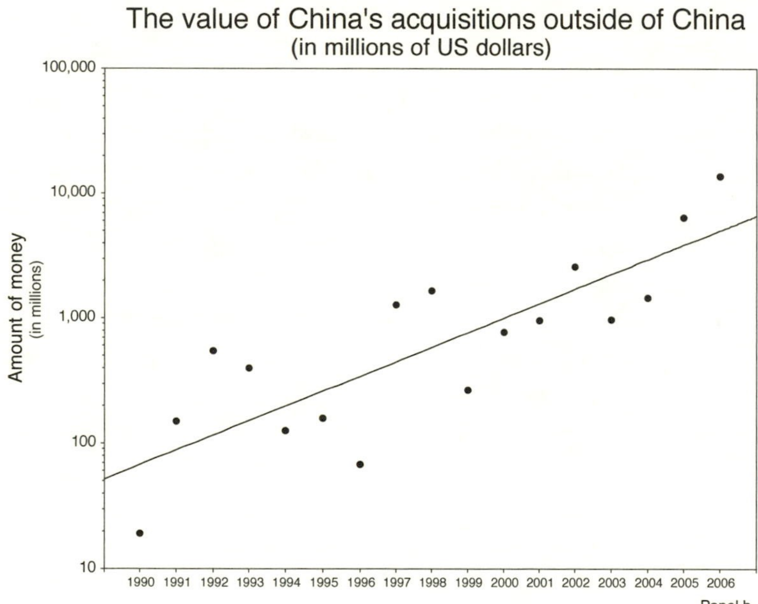


**Figure 11.14.** The data from figure 11.11 redrafted as two scatter plots. The plot of money is shown on a log scale, which linearizes the relationship between the amounts and the passing years. The superposition of regression lines on both panels allows the viewer to draw quantitative inferences about the rates of growth that was not possible with the depiction shown in figure 11.11.

all the quantitative information necessary to do these calculations, had a primarily qualitative message.

> *Magnum esse solem philosophus probabit, quantus sit mathematicus.**
> (Seneca, *Epistulae* 88.27)

### 11.5. CONCLUSION

William Playfair set a high standard for effective data display more than two hundred years ago. Since that time rules have been codified[3] and many books have been published that describe and exemplify good graphical practice.[4] All of these have had an effect on graphical practice. But it would appear from my sample of convenience that the effect was larger on the mass media than on the scientific literature. I don't know why, but I will postulate two possible reasons. First, because scientists make graphs with the software they have available and will tend, more often than is proper, to accept whatever are the default options for that software. Producers of displays for large market mass media have, I suspect, a greater budget and more flexibility. The second reason why poor graphical practices persist is akin to Einstein's observation of the persistence of incorrect scientific theory: "Old theories never die, just the people who believe in them."

* Roughly translated as "A philosopher will say that the sun is big, but a mathematician will measure it."

# 12

## Old Mother Hubbard and the United Nations

Wealth is not without its advantages and the case to the contrary, although it has often been made, has never proved widely persuasive.

—John Kenneth Galbraith, *The Affluent Society*, 1958

The science of uncertainty is often divided into two parts: exploration and confirmation. Most of the philosophy and tools of confirmatory statistics were developed over the course of the twentieth century. It is judicial in nature, where evidence is weighed and a decision is made. Exploratory methods are older, although the practice of exploratory data analysis was only given real scientific credence after the 1977 publication of John Tukey's remarkable book *Exploratory Data Analysis*. Exploratory analysis is analogous to detective work, in which clues are uncovered and followed, and from them hypotheses and plausible explanations are generated. The results of exploratory analyses form the grist that is subsequently subjected to the scrutiny required for confirmation. For the adventurer, the exploratory part is where the action is. This chapter is a case study that is meant as a simple introduction to exploratory analysis. In it a variety of tools are introduced in the pursuit of understanding.

Our tale of discovery begins innocently enough as a homework assignment in a statistics course for undergraduates at the University of Pennsylvania. Students were asked to find a publicly available data display that could be modified to serve its purpose better. One student found a table on the United Nations website[1] describing an aspect of the crowdedness of housing in sixty-three countries (table 12.1). The table is arranged alphabetically and contains the year that the data were gathered and the average number of persons per room in that country. This latter figure was also broken down into its rural and urban components.

This table adequately archived the information, but it was a long way from an evocative display to illuminate the character of housing crowdedness. The path toward revision and the subsequent enlightenment forms the subject matter of this chapter.

Revision was an iterative process of emails and conversations between the student and the instructor, which pointed toward reanalyses and supplemental data gathering. The first issue concerned the wide range of years during which the data were gathered. Did it make sense to compare Cameroon's crowdedness in 1976 with Brazil's in 1998? Two options were considered. The first was only to consider those countries whose data were no more than ten years old, which would have resulted in trimming off eleven countries. The second option was more empirical; it examined a plot of crowdedness against the year the data were gathered. The notion was that if there was a substantial overall trend (either increasing or decreasing) we could consider it as evidence that year mattered and we would indeed be erring if we considered all the data together. This, of course, assumes that when the United Nations chose to gather data was not dependent on how crowded that country was thought to be. The investigatory plot (figure 12.1) indicated no such trend.

Once this evidence was in hand, the next question that presented itself was whether each country was well characterized by its total crowdedness, or whether it was important to preserve, at least initially, the rural and urban breakdown. It seemed plausible to believe that there is likely to be more crowdedness in urban areas than rural ones, and hence countries that are less developed should have less crowdedness.

TABLE 12.1
**Average Number of Persons per Room (Excluding Bathrooms and Toilet Rooms)**

| *Country or area* | *Year* | *Total* | *Urban* | *Rural* |
|---|---|---|---|---|
| Aruba | 1991 | 0.7 | — | — |
| Austria | 1997 | 0.7 | 0.7 | 0.7 |
| Azerbaijan | 1998 | 2.1 | 1.9 | 2.3 |
| Bahamas | 1990 | 1.3 | 1.3 | 1.1 |
| Belgium | 1991 | 0.6 | — | 0.6 |
| Bermuda | 1991 | 0.6 | 0.6 | — |
| Bolivia | 1988 | — | 1.7 | — |
| Brazil | 1998 | 0.7 | 0.7 | 0.8 |
| Bulgaria | 1992 | 1.0 | 1.2 | 0.8 |
| Cameroon | 1976 | 1.2 | 1.2 | 1.3 |
| Canada | 1996 | 0.5 | 0.5 | 0.5 |
| China, Macao SAR | 1996 | 1.1 | — | — |
| Colombia | 1993 | 1.4 | 1.3 | 1.7 |
| Costa Rica | 1997 | 0.9 | 0.8 | 1.0 |
| Croatia | 1991 | 1.2 | — | — |
| Cuba | 1981 | 1.0 | 1.0 | 1.0 |
| Cyprus | 1992 | 0.6 | 0.6 | 0.7 |
| Czech Republic | 1991 | 1.0 | 1.1 | 1.0 |
| Egypt | 1996 | 1.3 | 1.3 | 1.4 |
| Finland | 1998 | 0.8 | 0.8 | 1.0 |
| France | 1990 | 0.7 | 0.7 | 0.7 |
| French Guiana | 1990 | 1.1 | 1.1 | 1.5 |
| Gambia | 1993 | 1.5 | 1.3 | 1.6 |
| Germany | 1987 | 0.5 | 0.5 | 0.5 |
| Guadeloupe | 1990 | 0.9 | 0.9 | 0.9 |
| Guam | 1990 | 0.8 | 0.7 | 0.8 |
| Honduras | 1988 | 2.2 | 1.8 | 2.6 |
| Hungary | 1990 | 0.8 | 0.8 | 0.8 |
| India | 1981 | 2.7 | 2.6 | 2.8 |
| Iraq | 1987 | 1.5 | 1.8 | 1.0 |
| Israel | 1983 | 1.2 | — | — |
| Japan | 1978 | 0.8 | 0.8 | 0.7 |
| Korea, Republic of | 1995 | 1.1 | 1.2 | 0.9 |
| Kuwait | 1985 | 1.7 | — | — |
| Lesotho | 1996 | 2.1 | — | — |
| Martinique | 1990 | 0.9 | 0.9 | 0.9 |

*(Continued)*

Table 12.1 (*Continued*)

| *Country or area* | *Year* | *Total* | *Urban* | *Rural* |
|---|---|---|---|---|
| Mexico | 1995 | — | 1.4 | — |
| Netherlands | 1998 | 0.7 | 0.6 | 0.6 |
| New Caledonia | 1989 | 1.2 | 1.1 | 1.4 |
| New Zealand | 1991 | 0.5 | 0.5 | 0.5 |
| Nicaragua | 1995 | 2.6 | 2.2 | 3.1 |
| Norway | 1990 | 0.6 | 0.6 | 0.6 |
| Pakistan | 1998 | 3.0 | 2.8 | 3.1 |
| Panama | 1990 | 1.6 | 1.4 | 1.9 |
| Peru | 1990 | 2.0 | 1.9 | 2.4 |
| Poland | 1995 | 1.0 | 0.9 | 1.1 |
| Portugal | 1991 | 0.7 | 0.7 | 0.7 |
| Puerto Rico | 1990 | 0.7 | 0.6 | 0.7 |
| Reunion | 1990 | 1.0 | 1.0 | 0.9 |
| Romania | 1992 | 1.3 | 1.3 | 1.2 |
| San Marino | 1991 | 0.7 | 0.7 | 0.8 |
| Serbia and Montenegro | 1991 | 1.2 | 3.1 | 3.3 |
| Slovakia | 1991 | 1.2 | 1.2 | 1.2 |
| Sri Lanka | 1981 | 2.2 | 2.3 | 2.1 |
| Sweden | 1990 | 0.5 | 0.5 | 0.5 |
| Switzerland | 1990 | 0.6 | 0.6 | 0.6 |
| Syrian Arab Republic | 1994 | 2.0 | 1.8 | 2.3 |
| Turkey | 1994 | 1.3 | 1.3 | 1.4 |
| United Kingdom | 1996 | 0.5 | 0.5 | 0.4 |
| United States | 1997 | 0.5 | 0.5 | 0.5 |
| U.S. Virgin Islands | 1990 | 0.6 | — | — |
| Uruguay | 1996 | 1.0 | 1.0 | 1.0 |

Happily, this too was subject to empirical verification. So the student made a plot (figure 12.2) comparing each country's rural with its urban crowdedness. The idea was that if they were highly related we would do no harm characterizing a country by its total figure. What we discovered was that, although rural and urban crowding is, in fact, strongly related, there was a tendency for rural areas to be more crowded than urban ones. This surprising result was an important clue for subsequent investigations.

Once these preliminaries were out of the way, the student felt justified in devising an alternative display showing the total crowdedness of

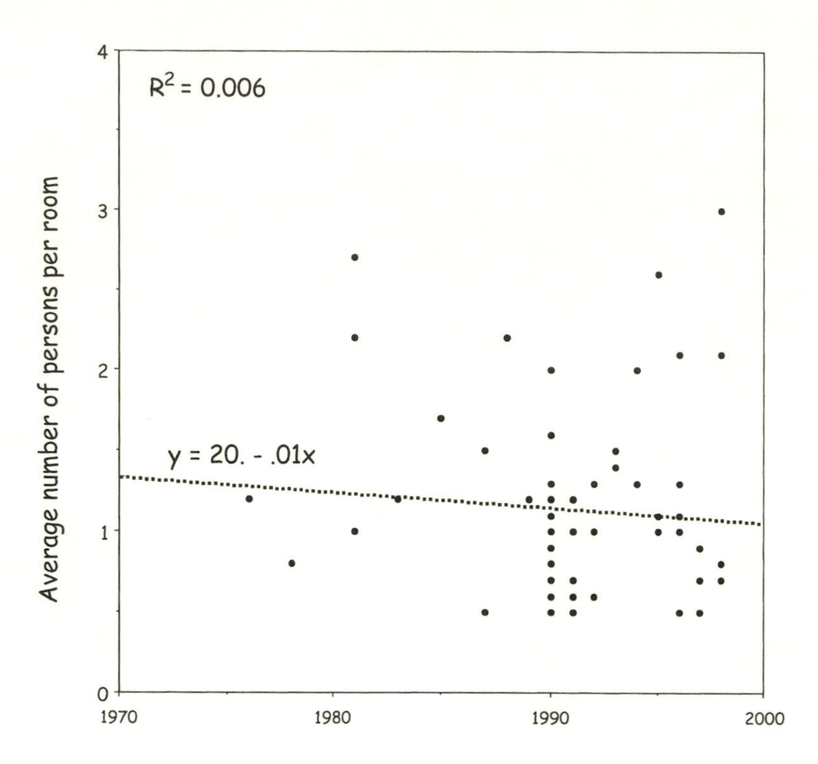


**Figure 12.1.**
There is essentially no relationship between housing crowdedness and the year the data were collected.

all the countries, knowing that the structure observed was unlikely to misrepresent any country. The goal of the display would be to show the distribution of crowdedness as well as to provide a tentative grouping of countries on this important variable. The display developed (figure 12.3) was ordered by crowdedness, not the alphabet, and provides an evocative view.

It was natural to look at the order of the countries in figure 12.3 and ask if there was some underlying variable that might "explain" why some nations were so much more crowded than others. The results in figure 12.2 allowed us to rule out urban/rural as a major explanation; what else might it be? It seemed obvious that wealth was a plausible differential explanation, since the richer nations tended to collect at the low end of the crowdedness continuum and the poor nations were at the other. The same United Nations website yielded a table of per capita gross domestic product (GDP) for the same set of countries; a plot

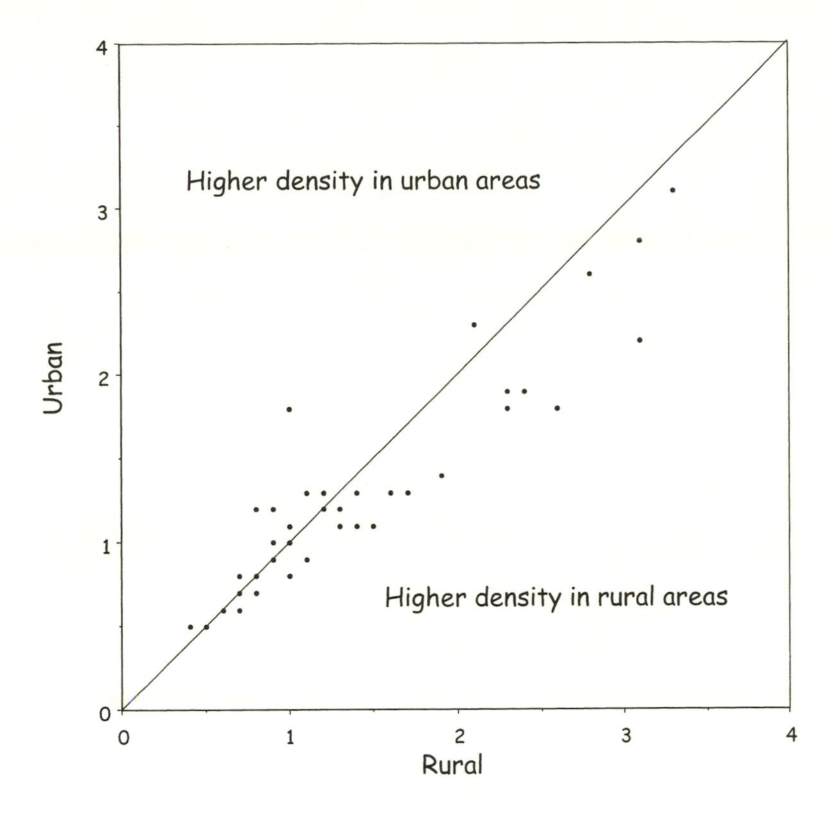


Figure 12.2.
Although there is a high correlation between urban and rural housing density, there is a tendency for more people per room in rural areas than urban.

of GDP vs. crowdedness (figure 12.4) showed a strong inverse relationship between the two; as GDP increased, crowdedness decreased.

But the metric that the data are collected in is not always the most informative way to view them. When this happens the reexpression of variables can help. The pileup of data points at the low end of GDP screamed for reexpression; some way to stretch out the data so that their character could be seen more clearly. A box-and-whisker plot* of GDP (figure 12.5) clearly shows a long straggly tail upwards. This sort

* This kind of plot is meant to show the shape of the distribution of the data. Its key elements are a box that contains the middle 50% of the data adorned with a line inside it that designates the median, and two whiskers emanating from its top and bottom that represent the top 25% and the bottom 25% respectively. Figure 12.5 has a few other features, of lesser importance for this account, that indicate an outlier and standard error bounds around the median.

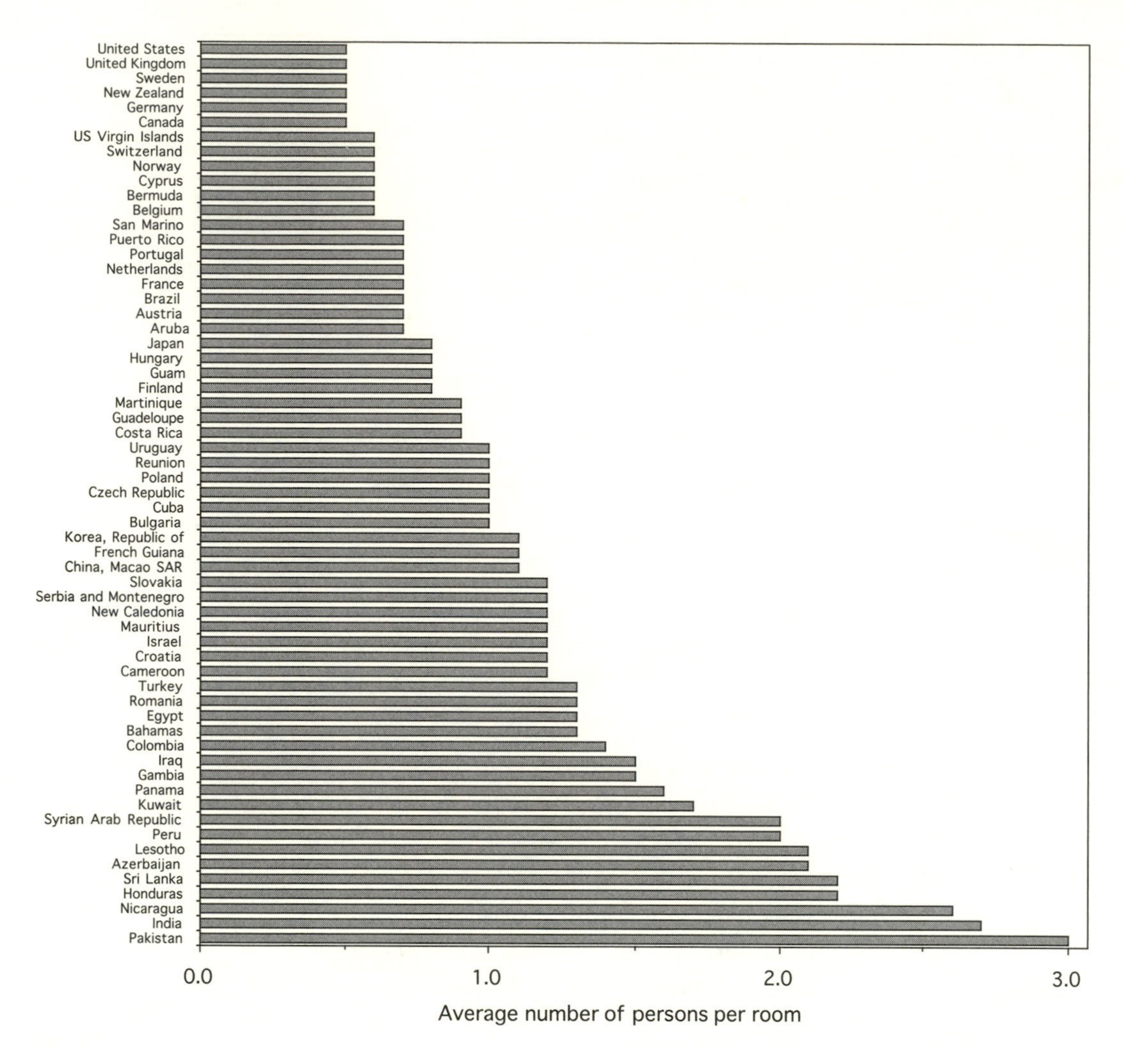


**Figure 12.3.**
U.N. Housing data indicate that the number of people per room varies by a factor of six over the sixty-two countries surveyed.

of shape is commonly transformed by taking logarithms. If we do so, we find that log(GDP) is a new scale that repairs the asymmetry, and allows us to return to other analyses that may help us to understand better the plausible causes of crowdedness (figure 12.6).

Now when crowdedness is plotted against log (GDP) (figure 12.7) a nonlinear relationship is observed. Unfortunately, while humans are good at spotting straight lines, we are less able to know what is the exact character of a curve when we spot one.

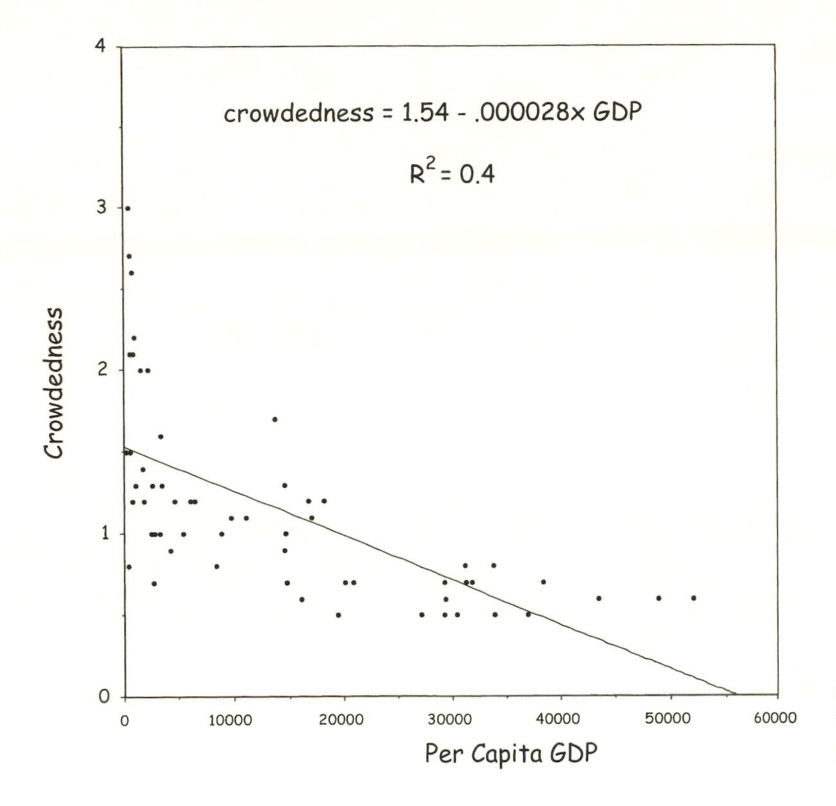


**Figure 12.4.**
There is a pileup of data at the low end of GDP.

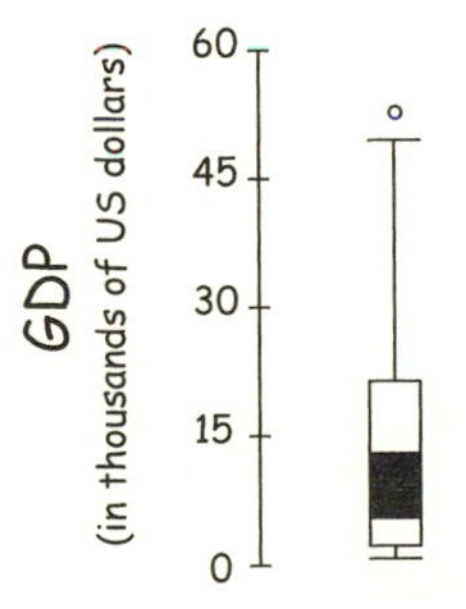


**Figure 12.5.**
Box-and-whisker plot of GDP for the sixty-three Countries shown in table 12.1.

Happily, Tukey[2] provided us with some tricks to help determine exactly what is the structure of any nonlinear function we spot in a data set. The basic idea is to try various transformations of the variable on the vertical axis until we find one that straightens the plot. A natural choice here would be either a square root transformation or, slightly

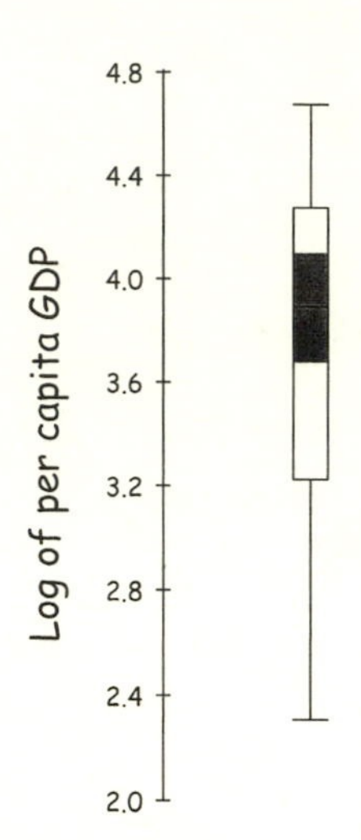


Figure 12.6.
Box-and-whisker plot of GDP data after transforming them by taking their logarithm.

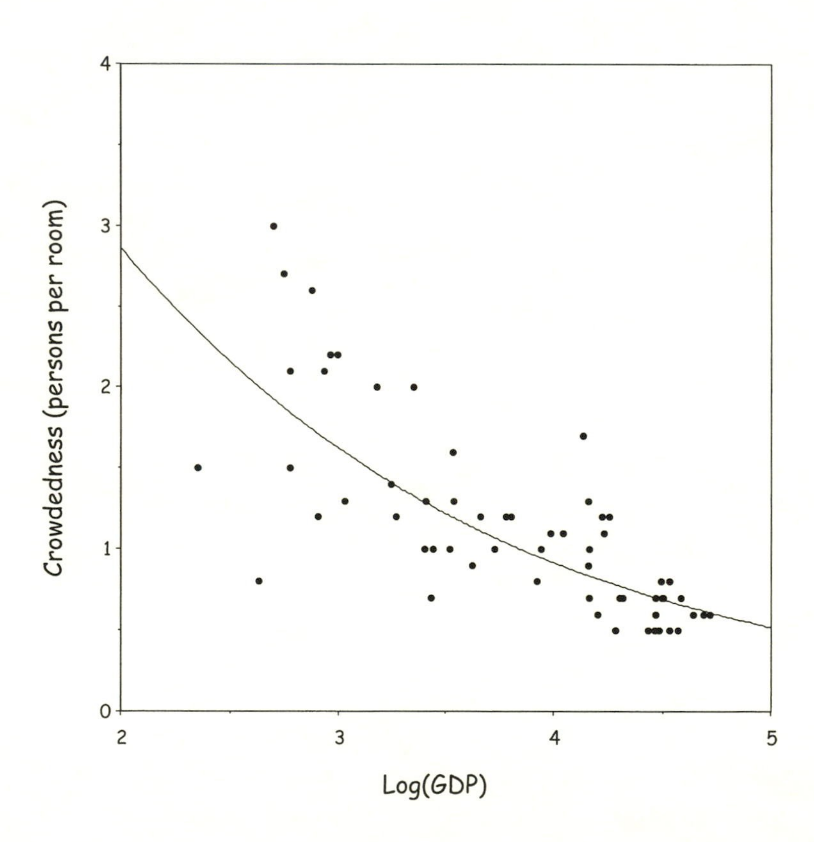


Figure 12.7.
Taking logs of GDP opens up the plot and reveals a nonlinearity.

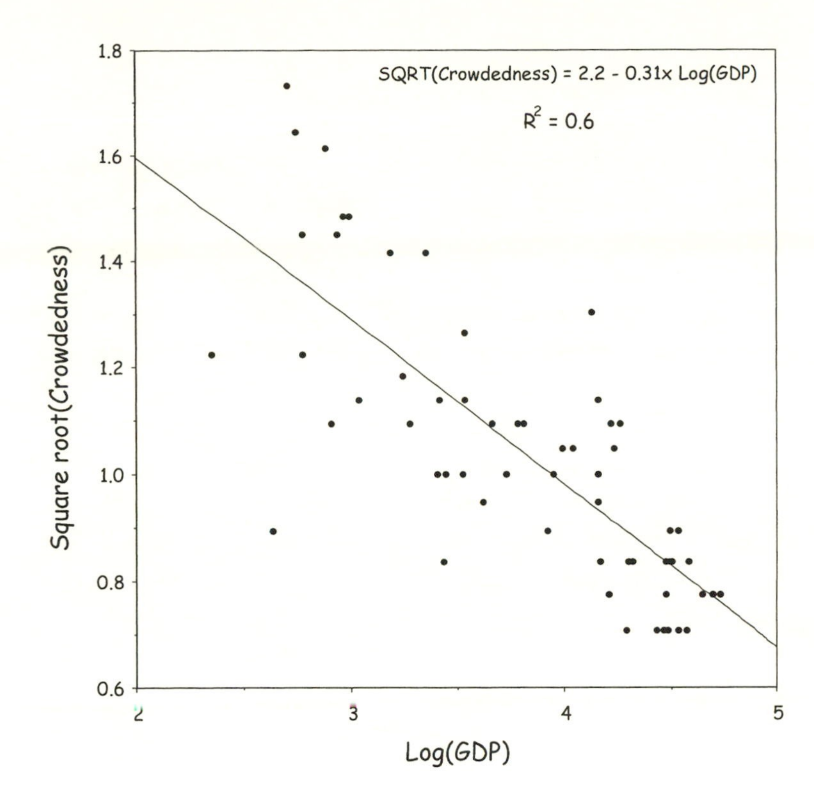


**Figure 12.8.**
The square root of crowdedness is linear in log(GDP).

more extreme, another logarithmic transformation. Trying the square root first and plotting the square root of crowdedness vs. log (GDP) (figure 12.8) seemed to provide a neat ending to this stage of the exploration. If we wanted to derive a formula to connect crowdedness to wealth it would be of the form

$$\text{SQRT (Crowdedness)} = 2.2 - 0.31 \times \log(\text{GDP}),$$

or

$$\text{Crowdedness} = [2.2 - 0.31 \times \log(\text{GDP})]^2. \qquad (12.1)$$

But this is not the end. The next step is to use equation (12.1) just derived to predict crowdedness and to subtract the predicted values from the actual ones. The results of this are called *residuals,** or more commonly, the errors we made in prediction. Often, if the residuals are plotted and looked at very hard, they can suggest another variable that might be responsible for them. It was with this goal in mind that we built a stem-and-leaf diagram of the residuals, using each country's name as the stem (figure 12.9).

Initially, we considered climate as an influence on crowdedness. It seemed plausible that countries with tropical climates, for example, might appear more crowded per room, given that a great deal of living is likely to be outdoors throughout the year, whereas in colder climates people are forced to stay indoors, and comfort would require more space. This possibility was quickly dismissed when we observed that tropical countries were seen on both ends of the distribution (Kuwait and India with large positive residuals, and Brazil, Gambia, and Cameroon with large negative ones). This confirmed that climate was unlikely to be a major explanatory factor.

A second possibility was that perhaps the structure of the residuals could be, at least partially, explained by the way that women were treated in each society; with countries that limited women's rights being more crowded than would be expected from their GDP. What might be an indicator variable that would capture this? Two obvious and available candidates were fertility rate and unemployment among women.

As we saw previously with the examination of GDP, the distribution of each variable needs to be plotted to see if some reexpression is necessary. In figure 12.10 are three box plots for fertility and two reexpressions of the data (square root and log). It is clear that the log transform was the most successful at making the distribution of the variable "fertility rate" symmetric. The same is true for unemployment.

* "Residuals" is technical jargon for what is left over after the prediction—the error. In this instance we were predicting the amount of crowdedness from the country's GDP (actually the log(GDP)). As is visible in the plot, we can do a pretty good job of predicting crowdedness from GDP, and the prediction equation, once we transform the two component variables, is a simple straight line. But the prediction, though pretty good, is a long way from perfect. And science, being the iterative enterprise that it is, demands that we try to do better. So we next looked at the residuals, the pieces of crowdedness not predicted from GDP, and tried to find another variable that would predict them.

| Residuals | Country/Region |
|---|---|
| 0.36 | Kuwait, Pakistan |
| 0.34 | |
| 0.32 | |
| 0.30 | Nicaragua |
| 0.28 | India |
| 0.26 | |
| 0.24 | Peru |
| 0.22 | Bahamas, Israel, New Caledonia |
| 0.20 | Honduras |
| 0.18 | Sri Lanka, Syria |
| 0.16 | |
| 0.14 | Azerbaijan, China, Macao, Panama |
| 0.12 | |
| 0.10 | Lesotho, Republic of Korea, Japan |
| 0.08 | Reunion, French Guiana |
| 0.06 | Croatia, Finland, Slovakia |
| 0.04 | San Marino, Guadeloupe, Martinique, Turkey, Mauritius, Netherlan |
| 0.02 | |
| 0.00 | Austria, Czech, France, Norway, Switzerland |
| -0.02 | Romania |
| -0.04 | Columbia, Puerto Rico |
| -0.06 | Aruba |
| -0.08 | Poland |
| -0.10 | Belgium, Portugal, |
| -0.12 | United States |
| -0.14 | Serbia, Montenegro, Sweden, Hungary, |
| -0.16 | United Kingdom |
| -0.18 | Urguay, Germany, Iraq, Egypt |
| -0.20 | Canada, Cyprus, Costa Rica, |
| -0.22 | Bulgaria |
| -0.24 | |
| -0.26 | New Zealand |
| -0.28 | |
| -0.30 | |
| -0.32 | Cameroon |
| -0.34 | |
| -0.36 | Gambia |
| -0.38 | |
| -0.40 | |
| -0.42 | |
| -0.44 | |
| -0.46 | |
| -0.48 | |
| -0.50 | |
| -0.52 | |
| -0.54 | |
| -0.56 | Brazil |

**Figure 12.9.** A stem-and-leaf diagram of the residuals in crowdedness after adjusting for differences in GDB.

Figure 12.10.
Three box-and-whisker plots of fertility rates for the sixty-three Countries. Each plot reflects the distribution after a different transformation, as indicated.

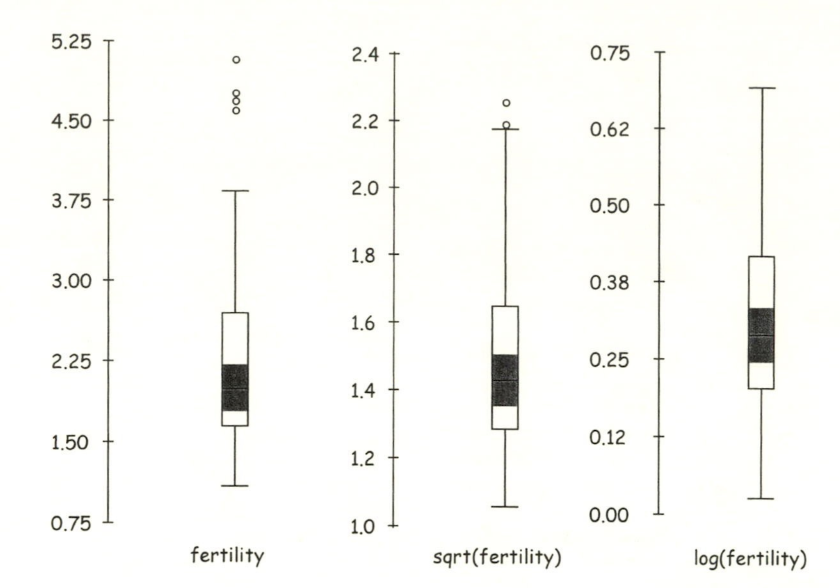


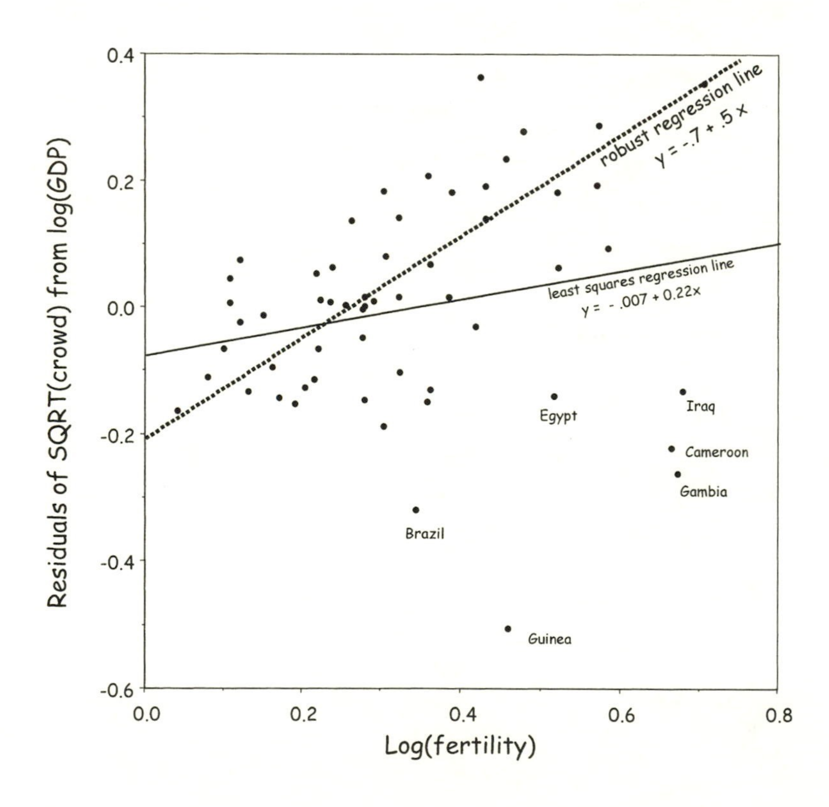


Figure 12.11.
Four highly influential points make the least squares regression undesirable, and a more robust version is preferred.

With the end of the trail seemingly in sight, a scatter plot was drawn showing the residuals against log(fertility) (figure 12.11). We fit two different straight lines in an effort to represent the linear relationship between the residuals of crowdedness (residuals from log[GDP]) and the log(fertility). One line was fitted by minimizing the sum of the squares of the errors from the line (a least squares fit) and the other was done in another way that is less severely affected by the indicated outliers.

The least squares line (shown as the solid line in the plot) was clearly affected by a few influential points, and a more robust alternative (the dashed line) was considered as a better representation of the structure of the relationship between the two variables.

We can summarize progress up to this point with an equation and a graph. The equation is

$$\text{Square root (crowdedness)} = 1.85 - .24 \log(\text{GDP}) + .38 \log(\text{fertility}).$$

This equation can be interpreted qualitatively in the not very surprising way that a country's housing becomes more crowded as the wealth of the country decreases and the fertility rate of its women increases.

A helpful graph is a plot of the residuals from this model (figure 12.12) indicates that Mexico is less crowded than this model would suggest (as are, to a lesser extent, Guinea, Gambia, Brazil, and Cameroon), whereas Kuwait, India, and Pakistan are more crowded than the model predicts. Further investigations showed that these residuals are shrunken still further when log(unemployment among women) is added to the model, but the lessons learned from this aspect of the analysis do not justify the space that its inclusion would require.

The preceding narrative was meant to show how the search to answer a small question can grow and thus provide an example of how statistical thinking helps us to understand our environment. It also shows how each step is simple, reinforcing our contention that the science of uncertainty is one that resides closer to the altitude where we all live and can be used broadly to help us navigate our uncertain world.

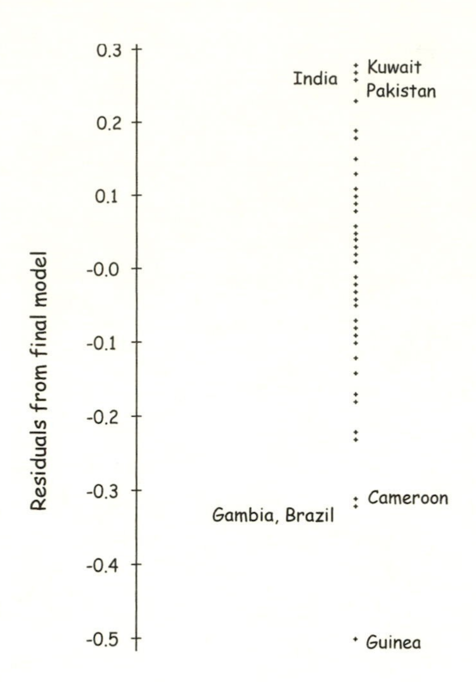


**Figure 12.12.**
A dot plot showing the residuals in crowdedness after adjusting for GDP and fertility.

It also illustrates the omnipresent, existential issue "how much is enough?" We ended this investigation with three independent variables (only two of which we reported). We could have gone further. When are you done? Doing exercises like this one helps to teach us the important lesson that, in any serious investigation, you are never done, you just run out of time—*obesa cantavit.**

* The fat lady has sung.

13

# Depicting Error*

## 13.1. INTRODUCTION

> One hallmark of the statistically conscious investigator is his firm belief that however the survey, experiment, or observational program actually turned out, it could have turned out somewhat differently. Holding such a belief and taking appropriate actions make effective use of data possible. We need not always ask explicitly "How much differently?" but we should be aware of such questions. Most of us find uncertainty uncomfortable . . . (but) . . . each of us who deals with the analysis and interpretation of data must learn to cope with uncertainty.
>
> —Mosteller and Tukey, 1968, p. 100

Whenever we discuss information we must also discuss its accuracy. Effective display of data must

(1) remind us that the data being displayed do contain some uncertainty, and then
(2) characterize the size of that uncertainty as it pertains to the inferences we have in mind, and in so doing
(3) help keep us from drawing incorrect conclusions through the lack of a full appreciation of the precision of our knowledge.

*This chapter has some technical aspects that are hard to avoid; if you have never taken a statistics course, or need to be reminded of some of the concepts, there is a brief tutorial at the end of this chapter. You can either read it first, or turn to it as required, when you run across a term or concept that is unfamiliar.

The examples chosen here focus principally on errors in means, but the graphical ideas expressed should generalize easily to other situations. Throughout this discussion we assume that estimates of precision are available, and thus the task is strictly one of effective display. This assumption is admittedly a big one. Great efforts and much imagination have been expended in the search for the true uncertainty. Modern statistics has gone far beyond characterizing error by dividing the observed standard deviation by $\sqrt{n}$ (remember the most dangerous equation discussed in chapter 1). Ingenious methods that require intensive computing have been developed over the past thirty years or so (sensitivity analysis through resampling or multiple imputation) provide some inroads into the measurement of the real uncertainty. Despite my appreciation for the importance of this work, I will ignore how the uncertainty is calculated and proceed assuming that, in whatever way was required, such estimates are available and our task is simply to convey them as well as we can.

## 13.2. ERRORS IN TABLES

Anyone can calculate a mean, it takes a statistician to compute a variance.

—Paul Holland, 2000

Let us begin our examination of the depiction of error with a typical tabular display. Shown in table 13.1 is a data table extracted from a much larger display. Omitted are thirty states and two territories as well as another variable (percentage in each category) and its standard error. As abbreviated, this table still contains the elements important to our discussion, without being unwieldy. The only deviation from standard format in this table is that the standard error of each data entry is in a separate column rather than denoted within parentheses adjacent to the data entry whose variability it characterizes. This was done primarily to ease text manipulation. There are three lines added to the bottom of this table. These lines are the point of this discussion.

Showing the standard errors of statistics in this way certainly satisfies requirement (1). The visual weight given to the numbers connoting error is the same as that which conveys the data. This depiction does not, however, allow us to summarize easily the structure of error.

Table 13.1
**Average Mathematics Proficiency by Parents' Highest Level of Education, Grade 8, 1992**

| | *Graduated college* | | *Some education after high school* | | *Graduated high school* | | *Did not finish high school* | | *I don't know* | |
|---|---|---|---|---|---|---|---|---|---|---|
| *Public schools* | $\theta$ | *se* | $\theta$ | *se* | $\theta$ | *se* | $\theta$ | *se* | $\theta$ | *se* |
| **NATION** | **279** | **1.4** | **270** | **1.2** | **256** | **1.4** | **248** | **1.8** | **251** | **1.7** |
| **STATES** | | | | | | | | | | |
| Alabama | 261 | 2.5 | 258 | 2.0 | 244 | 1.8 | 239 | 2.0 | 237 | 2.9 |
| Arizona | 277 | 1.5 | 270 | 1.5 | 256 | 1.6 | 245 | 2.5 | 248 | 2.7 |
| Arkansas | 264 | 1.9 | 264 | 1.7 | 248 | 1.6 | 246 | 2.4 | 245 | 2.7 |
| California | 275 | 2.0 | 266 | 2.1 | 251 | 2.1 | 241 | 2.2 | 240 | 2.9 |
| Colorado | 282 | 1.3 | 276 | 1.6 | 260 | 1.5 | 250 | 2.4 | 252 | 2.6 |
| Connecticut | 288 | 1.0 | 272 | 1.8 | 260 | 1.8 | 245 | 3.3 | 251 | 2.4 |
| Delaware | 274 | 1.3 | 268 | 2.3 | 251 | 1.7 | 248 | 4.0 | 248 | 3.4 |
| District of Columbia | 244 | 1.7 | 240 | 1.9 | 224 | 1.6 | 225 | 3.2 | 229 | 2.2 |
| Florida | 268 | 1.9 | 266 | 1.9 | 251 | 1.8 | 244 | 2.7 | 244 | 3.2 |
| Georgia | 271 | 2.1 | 264 | 1.7 | 250 | 1.3 | 244 | 2.2 | 245 | 2.6 |
| Hawaii | 267 | 1.5 | 266 | 1.9 | 246 | 1.8 | 242 | 3.5 | 246 | 2.1 |
| Idaho | 281 | 0.9 | 278 | 1.3 | 268 | 1.4 | 254 | 2.3 | 254 | 2.8 |
| | · | · | · | · | · | · | · | · | · | · |
| | · | · | · | · | · | · | · | · | · | · |
| | · | · | · | · | · | · | · | · | · | · |

*Error terms for comparison*

| | | | | | |
|---|---|---|---|---|---|
| Max Std error of difference | 3.5 | 3.4 | 3.5 | 6.4 | 6.4 |
| 40 Bonferroni (std. err × 3.2) | 11.3 | 11.0 | 11.3 | 20.7 | 20.7 |
| 820 Bonferroni (std. err × 4.0) | 14.0 | 13.6 | 14.0 | 25.6 | 25.6 |

$\theta$ is an estimate of the average proficiency in the state.

Source: Abstracted from Table 2.12 in *The 1992 NAEP Trial State Assessment.*

For example, we cannot tell if the size of the error is related to the proficiency, nor is it easy to see how much variation there is in the error over the various states or across the categories of parental education. The answers to these kinds of questions are important for requirement (2).

In this case, errors across states are quite homogeneous, and so displaying the maximum value of the standard error as a summary allows us to extract a handy, conservative value from the error terms provided. Using such a summary term (rather than, say, a mean or a median) will lower the likelihood of our declaring states different when there is reasonable evidence that they might not be. This raises the second part of requirement (2) "uncertainty as it pertains to the inferences we have in mind." What kinds of question are likely to be used to query table 13.1?

When national data are reported broken down by states it is natural to assume that the disaggregated data are meant to be compared. Thus the standard errors are merely the building blocks that are required to be able to construct the standard error of the difference so that the statistical significance of the observed difference between state means can be ascertained. An upper bound on this standard error is obtained by multiplying the maximum standard error by the $\sqrt{2}$. This result is reported on the first line of the section at the bottom of the table labeled "Error terms for comparisons." It is a handy rule of thumb for anyone wishing to make a single comparison. Thus, suppose, for some reason, we are interested in comparing the performance of Hawaii's eighth graders whose parents were college graduates (267), with their counterparts in Delaware (274). We note that their mean scores differ by 7. This is twice the maximum standard error of the difference of the means (3.5). We can thus conclude that this observed difference is statistically significant beyond the nominal ($\alpha = .05$) value. We could have done this comparison for any single pair that was of special interest to us.

Although the comparison of a particular pair of states may be an occasional use for a data table like this, we suspect that comparing one state (perhaps our own) with all others is a more likely use. To do this correctly we need to control the artifactual increases in the error rate due to making many comparisons. The most common way to assure that the type I error rate is controlled is to boost the size of the standard error sufficiently that the overall error rate remains at the nominal (.05) level. In the original table there were forty-one states, and so comparing any one of these with each of the others would yield a total of forty separate *t*-tests. The Bonferroni inequality[1] provides us with a conservative rule for combining the standard error of the difference with the forty different $t$-tests to yield the ordinate associated with a critical region of the right size for the entire family of tests. This figure is provided on the line labeled "40 Bonferroni." Thus if we wish to compare Hawaii's performance (among children whose parents were college graduates) with that of each of the other forty states that participated in the state assessment, we can declare any differences greater than 11.3 points "statistically significant beyond the .05 level." The 11.3 point decision rule is conservative (for most comparisons about 50% too large), but it provides a safe and quick rule of thumb.

If someone is interested in comparing each state with all others there will be 820 comparisons (the number of different ways we can choose two states from among forty one). To control the type I error rate in this situation requires boosting the critical region still further (to 14.0). Such a figure is given next to the label "820 Bonferroni." We are uncertain of the usefulness of including such a figure, although it still works, because anyone who is really interested in making all 820 comparisons will probably want a somewhat less conservative decision rule. Such a user of the data table would need to go back to the original standard errors and compute a more precise figure. For this particular data set, even though the standard errors are all reasonably similar, a 30–60 percent shrinkage in the decision rule will occur if the individual standard errors are used. Since our purpose here is not to explore alternative schemes for multiple comparisons we will not dwell on this aspect too much longer. However, depending on the circumstance, it may be profitable to choose a less conservative summary.*

A completely revised version of the original table[2] is shown as table 13.2. This table replaces the standard error columns completely with the summary standard errors. We have also reordered the rows of the table by the overall state performance,† inserted other summary statistics for comparison, spaced the table according to the data, and

*For example, Benjamini and Hochberg's FDR method (1995). The interested reader is referred to Williams, Jones, and Tukey (1994) for a much fuller exploration of the issues surrounding adjustment procedures for multiple comparisons within the National Assessment of Educational Progress (NAEP).

†The numbers alongside each state name are location aids. There is a separate locator table, alphabetical by state name, which provides these numbers for easy lookup. The idea of including an alphabetical locator table is common on maps ("Albany M6"), but has been in use for large tables for at least a century (e.g., Francis Walker, the director of the 1890 census, used it frequently in the tables describing the growth of cities in the 1890 census.

Table 13.2
**Average Mathematics Proficiency by Parents' Highest Level of Education, Grade 8, 1992**

| *Public schools* | *Graduated college* | *Some education after high school* | *Graduated high school* | *Did not finish high school* | *I don't know* | *Mean* |
|---|---|---|---|---|---|---|
| **NATION** | **279** | **270** | **256** | **248** | **251** | **267** |
| **STATES** | | | | | | |
| 1 Iowa | 291 | 285 | 273 | 262 | 266 | **283** |
| 2 North Dakota | 289 | 283 | 271 | 259 | 272 | **283** |
| 3 Minnesota | 290 | 284 | 270 | 256 | 268 | **282** |
| 4 Maine | 288 | 281 | 267 | 259 | 266 | **278** |
| 5 Wisconsin | 287 | 282 | 270 | 254 | 255 | **278** |
| 6 New Hampshire | 287 | 280 | 267 | 259 | 262 | **278** |
| 7 Nebraska | 287 | 280 | 267 | **247–** | 256 | **277** |
| 8 Idaho | 281 | 278 | 268 | 254 | 254 | **274** |
| 9 Wyoming | 281 | 278 | 266 | 258 | 260 | **274** |
| 10 Utah | 280 | 278 | 258 | 254 | 258 | **274** |
| 11 Connecticut | 288 | 272 | 260 | **245–** | 251 | **273** |
| 12 Colorado | 282 | 276 | 260 | 250 | 252 | **272** |
| 13 Massachusetts | 284 | 272 | 261 | 248 | **248–** | **272** |
| 14 New Jersey | 283 | 275 | 259 | 253 | 250 | **271** |
| 15 Pennsylvania | 282 | 274 | 262 | 252 | 252 | **271** |
| 16 Missouri | 280 | 275 | 264 | 254 | 252 | **271** |
| 17 Indiana | 283 | 275 | 260 | 250 | 249 | **269** |
| 18 Ohio | 279 | 272 | 260 | 243 | 249 | **268** |
| 19 Oklahoma | 277 | 272 | 257 | 254 | 251 | **267** |
| 20 Virginia | 282 | 270 | 252 | 248 | 251 | **267** |
| 21 Michigan | 277 | 271 | 257 | 249 | 248 | **267** |
| 22 New York | 277 | 271 | 256 | 243 | **240–** | **265** |
| 23 Rhode Island | 276 | 271 | 256 | 244 | **239–** | **265** |
| 24 Arizona | 277 | 270 | 256 | 245 | 248 | **264** |
| 25 Maryland | 278 | 266 | 250 | 240 | 245 | **264** |
| 26 Texas | 281 | 272 | 253 | 247 | 244 | **264** |
| 27 Delaware | 274 | 268 | 251 | 248 | 248 | **262** |
| 28 Kentucky | 278 | 267 | 254 | 246 | 242 | **261** |
| 29 California | 275 | 266 | 251 | 241 | 240 | **260** |

*(Continued)*

TABLE 13.2 (*Continued*)

| *Public schools* | *Graduated college* | *Some education after high school* | *Graduated high school* | *Did not finish high school* | *I don't know* | *Mean* |
|---|---|---|---|---|---|---|
| 30 South Carolina | 272 | 268 | 248 | 248 | 247 | **260** |
| 31 Florida | 268 | 266 | 251 | 244 | 244 | **259** |
| 32 Georgia | 271 | 264 | 250 | 244 | 245 | **259** |
| 33 New Mexico | 272 | 264 | 249 | 244 | 245 | **259** |
| 34 Tennessee | 267 | 265 | 251 | 245 | 243 | **258** |
| 35 West Virginia | 270 | **269 +** | 251 | 244 | 239 | **258** |
| 36 North Carolina | 271 | 265 | 246 | 240 | 240 | **258** |
| 37 Hawaii | 267 | 266 | 246 | 242 | 246 | **257** |
| 38 Arkansas | 264 | 264 | 248 | **246 +** | 245 | **256** |
| 39 Alabama | 261 | 258 | 244 | 239 | 237 | **251** |
| 40 Louisiana | 256 | 259 | 242 | 237 | 236 | **249** |
| 41 Mississippi | 254 | 256 | 239 | 234 | 231 | **246** |
| **Means** | **279** | **270** | **256** | **248** | **251** | **267** |
| **TERRITORIES** | | | | | | |
| 42 Guam | 246 | 244 | 229 | 224 | 226 | **235** |
| 43 District of Columbia | 244 | 240 | 224 | 225 | 229 | **234** |
| 44 Virgin Islands | 224 | 232 | 221 | 219 | 217 | **222** |
| | | | *Error terms for comparison* | | | |
| Max Std error of diff. | 3.5 | 3.4 | 3.5 | 6.4 | 6.4 | 3.5 |
| 40 Bonferroni | 11.3 | 11.0 | 11.3 | 20.7 | 20.7 | 11.3 |
| 820 Bonferroni | 14.0 | 13.6 | 14.0 | 25.6 | 25.6 | 14.0 |

indicated unusual data values.* Although a description of the value of these other changes is outside the purview of this chapter, we include this complete revision as an indication of how the revised error terms fit into a broader presentation scheme.[3]

We have learned several lessons from this tabular experiment. First, that including the standard errors of every data entry clogs the display, hindering our vision of the primary data structure. Second, that the

*"Unusual" in this context refers to a data entry whose residual from an additive model was, out of the 220 residuals, among the seven most extreme. These seven are typographically denoted with shading and a + or − indicating the direction of the residual.

inclusion of this additional complexity may not be especially helpful, even for the job the standard errors are intended for. Third, that a little thought about the prospective use of the data contained in the table may both allow a considerable simplification and increase the ease of use. And finally, fourth, that through this consideration we can provide protection for the user of the table (in this case through the explicit inclusion of the stepped-up confidence bounds for multiple comparisons) against the naive use of the standard errors. Thus the implementation demonstrated in table 13.2 satisfies all three of the desiderata of effective display of error. In the next section we will explore how these lessons translate into other display formats.

## 13.3. ERRORS IN GRAPHS

In figure 13.1 is a typical display that includes both data and associated error. The data are shown as large black dots; the errors are shown as vertical bars that are one standard error long in each direction. This is as typical as the tabular form of including standard errors as parenthetical additions to the data entries.

There are analogous versions of figure 13.1 for other graphical formats. In figures 13.2 and 13.3 are two (now outdated) approaches to adding a standard error bar to bar charts. Of course we need not fixate on just one standard error (which under the usual Gaussian assumptions corresponds to a 68% confidence bound on the mean); we could show one and a half or two standard errors(corresponding to 95% and 99% confidence bounds). In figure 13.4 are four versions of how this might be done. These are all traditional usages despite all of them being wasteful of space. It is certainly profligate to use an entire bar when all of the information about the mean is contained in the location of the top line; the rest is chartjunk (using Tufte's apt neologism).

### An Aside on Accurate Labeling

It is important to explain exactly what the bounds depicted are bounding. In each of the cases illustrated so far they are confidence bounds on the mean of the distribution. In the past there have been clear errors of understanding on this point. As an example, consider the graph in figure 13.5 which was mislabeled by its producers and

*(Continued)*

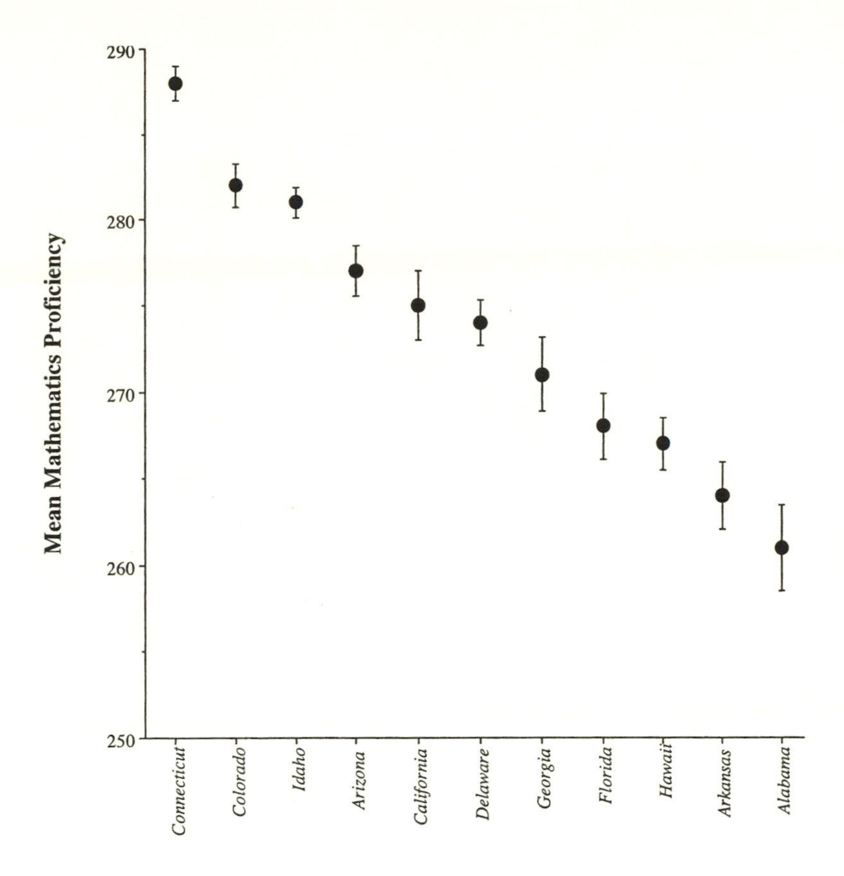


**Figure 13.1.**
Average mathematics proficiency for eighth grade children with at least one parent who graduated from college (1992 NAEP State Assessment data). A typical depiction of data including both their level and error. The large dots locate the position of the mean, the bars depict one standard error of the mean in each direction.

*(Continued)*
subsequently described erroneously by Calvin Schmid in his otherwise accurate reportage. Clearly the bounds specified are far too large to be confidence bounds on the mean. Neither could they be an accurate depiction of the distribution of years of schooling. Obviously they are showing the means with horizontal lines and using boxes and bars to denote one and two standard deviation bounds. This would represent the actual distribution of education only if that distribution was approximately Gaussian. We suspect that this is unlikely; that it is probable that at least at the turn of the century the distribution was skewed with a long straggly tail upward. How this skewness changed over time (if it did) would be an interesting question, unanswerable from this depiction.

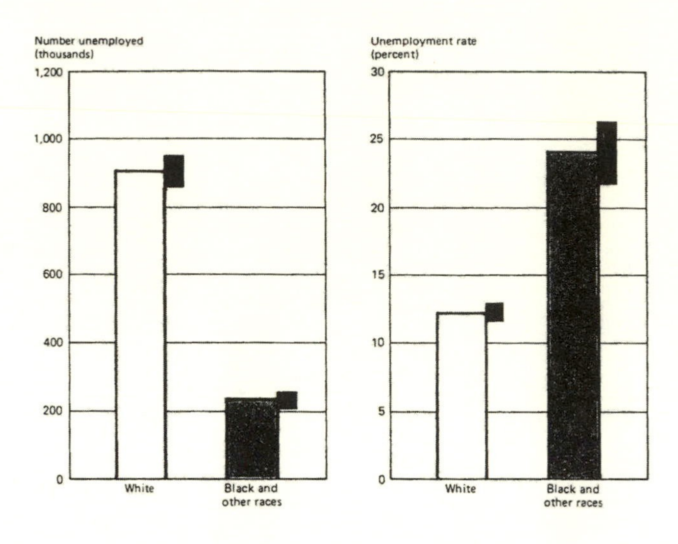


**Figure 13.2.**
A traditional method of depicting error on a bar chart using an error rectangle on the side of the bar. (From Office of Federal Statistical Policy and Standards, U.S. Bureau of the Census, *Social Indicators, 1976*, Washington, D.C.: Government Printing Office, 1977, p. xxvii.)

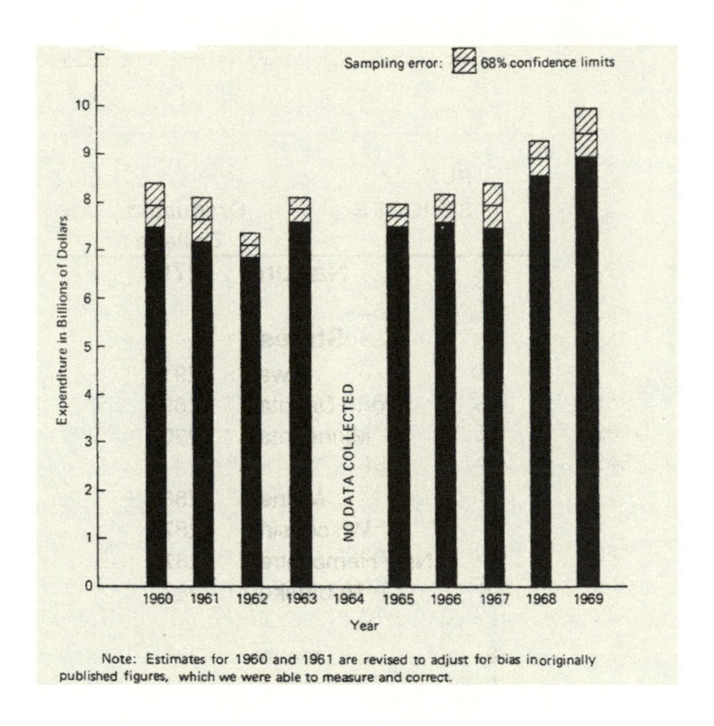


**Figure 13.3.**
A traditional method of depicting error on a bar chart using an error rectangle superimposed over the end of the bar. (From Maria E. Gonzalez et al., *Standards for Discussion and Presentation of Errors in Data*, Technical Paper 32, U.S. Bureau of the Census, Washington, D.C.: Government Printing Office, 1974, p. v–3.)

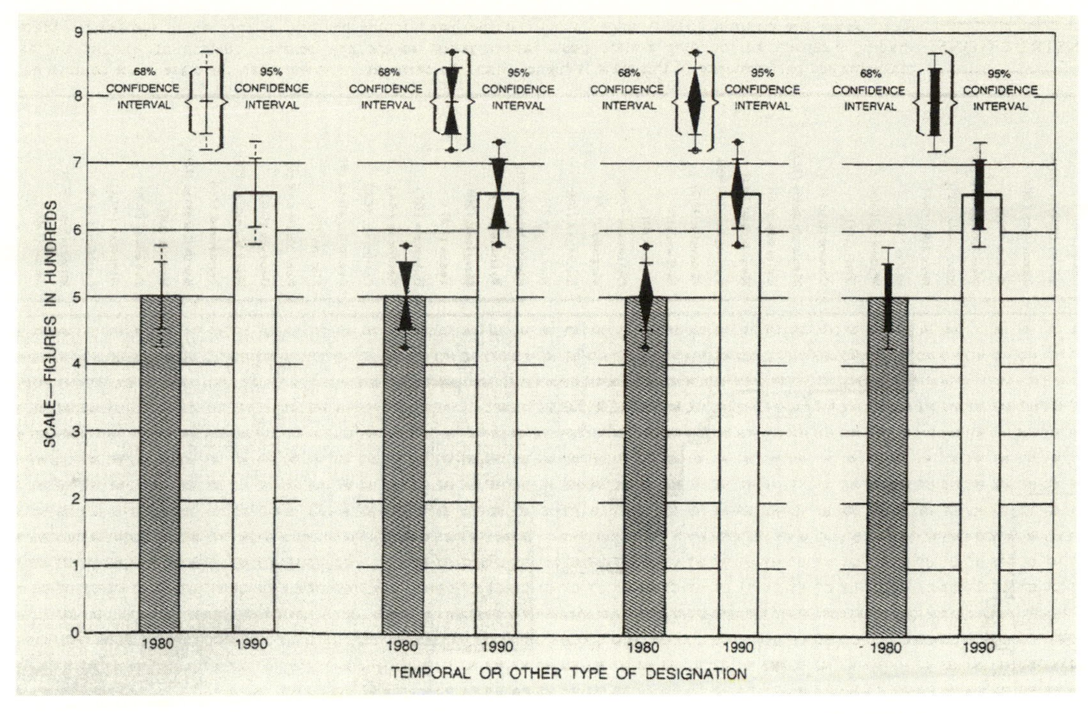


**Figure 13.4.**
Figure 10-3 from Schmid (1983, p. 195) showing four methods of depicting confidence bounds on a bar chart using various figurations superimposed over the end of the bar. Reprinted with permission.

An obvious modification to these figures would be to plot something analogous to the Bonferroni bounds, or perhaps half a bound. In the former case, anytime a point falls within such a bound from another point the two are not significantly different. In the latter, any overlap of the bars sticking out could be interpreted as "not significant." This sort of idea has been ingeniously implemented in the NAEP charts (whose resemblance to those found on panty-hose packages, has not gone unnoticed; see figure 13.6) in which only states whose bounds do not overlap are different enough in performance for that difference to be thought of as consequential.

In figure 13.7 is a simpler version of this same idea in which the error bars shown represent the 820 Bonferroni bounds. To compare any two states you have only to note if one state's point is outside the other's bounds. An alternative display to this (shown in figure 13.8) is one in which the error bars are half of these. It appears somewhat less cluttered and comparisons require noting whether two sets of bounds overlap. We have little intuition as to which of these two options is

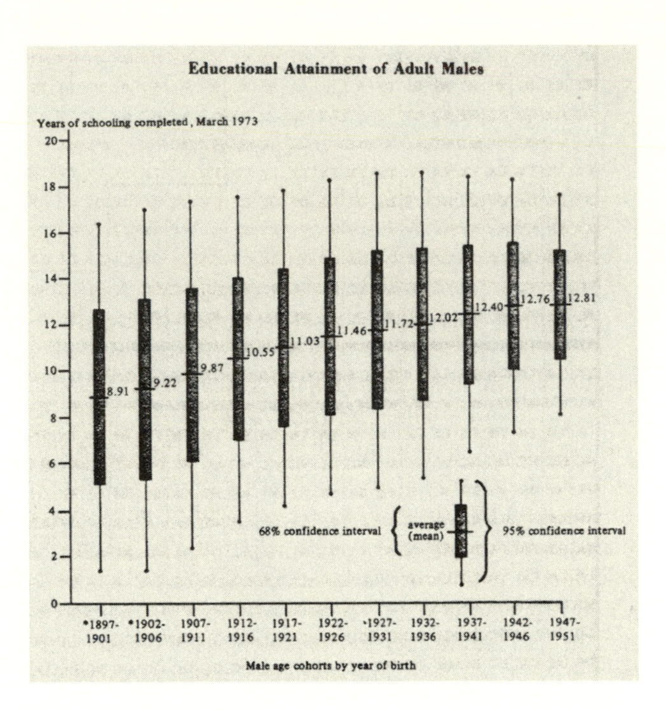


**Figure 13.5.**
A figure showing how incorrect labeling can lead to wildly inaccurate inferences. It reports confidence intervals on the mean, but in fact merely depicts the mean and error bars showing one and two standard deviations in each direction. (From Mary A. Golladay, *The Condition of Education, Part 1*, HEW National Center for Education Statistics, Washington, D.C.: Government Printing Office, 1977, p. 104.) It also misreports the source of the data which should probably be Table 1, p. 102 in Hauser and Featherman (1976).

preferred and thus will depart this aspect with no further recommendation except to try both and see what appeals to you.[4]

### 13.3.1. Changing the Metaphor

Making the physical size, and hence the visual impact, of data points proportional to their error with plotting conventions like error bars may be a bad idea. Not only does it often make the picture crowded, but it also gives greatest visual impact to the most poorly estimated points. Is this what we want to do?

For some uses, such a display metaphor may be useful. For example, if we wish to draw some sort of regression line through the points, we should try to get closer to a well-estimated point than to one that

Figure 13.6. A "pantyhose chart" that includes the Bonferroni expanded standard error of the mean within its bounds so that the user can scan down any column (state) and instantaneously decide which states are significantly different in their performance from that state. (Figure 1.6 from the *National Compendium for the NAEP 1992 Mathematics Assessment for the Nation and the States*, p. 40.)

INSTRUCTIONS: Read *down* the column directly under a state name listed in the heading at the top of the chart. Match the shading intensity surrounding a state postal abbreviation to the key below to determine whether the average reading performance of this state is higher than, the same as, or lower than the state in the column heading.

| New Hampshire (NH)* | Maine (ME)* | Massachusetts (MA) | North Dakota (ND) | Iowa (IA) | Wisconsin (WI) | Wyoming (WY) | New Jersey (NJ)* | Connecticut (CT) | Nebraska (NE)* | Indiana (IN) | Minnesota (MN) | Virginia (VA) | Pennsylvania (PA) | Utah (UT) | Oklahoma (OK) | Missouri (MO) | Idaho (ID) | Ohio (OH) | Rhode Island (RI) | Colorado (CO) | Michigan (MI) | West Virginia (WV) | New York (NY)* | Delaware (DE)* | Kentucky (KY) | Texas (TX) | Georgia (GA) | Tennessee (TN) | North Carolina (NC) | Maryland (MD) | Arkansas (AR) | New Mexico (NM) | South Carolina (SC) | Arizona (AZ) | Florida (FL) | Alabama (AL) | Louisiana (LA) | Hawaii (HI) | California (CA) | Mississippi (MS) | District of Columbia (DC) | Guam (GU) |
|---|---|---|---|---|---|---|---|---|---|---|---|---|---|---|---|---|---|---|---|---|---|---|---|---|---|---|---|---|---|---|---|---|---|---|---|---|---|---|---|---|---|---|
| NH | NH | NH | NH | NH | NH | NH | NH | NH | NH | NH | NH | NH | NH | NH | NH | NH | NH | NH | NH | NH | NH | NH | NH | NH | NH | NH | NH | NH | NH | NH | NH | NH | NH | NH | NH | NH | NH | NH | NH | NH | NH | NH |
| ME | ME | ME | ME | ME | ME | ME | ME | ME | ME | ME | ME | ME | ME | ME | ME | ME | ME | ME | ME | ME | ME | ME | ME | ME | ME | ME | ME | ME | ME | ME | ME | ME | ME | ME | ME | ME | ME | ME | ME | ME | ME | ME |
| MA | MA | MA | MA | MA | MA | MA | MA | MA | MA | MA | MA | MA | MA | MA | MA | MA | MA | MA | MA | MA | MA | MA | MA | MA | MA | MA | MA | MA | MA | MA | MA | MA | MA | MA | MA | MA | MA | MA | MA | MA | MA | MA |
| ND | ND | ND | ND | ND | ND | ND | ND | ND | ND | ND | ND | ND | ND | ND | ND | ND | ND | ND | ND | ND | ND | ND | ND | ND | ND | ND | ND | ND | ND | ND | ND | ND | ND | ND | ND | ND | ND | ND | ND | ND | ND | ND |
| IA | IA | IA | IA | IA | IA | IA | IA | IA | IA | IA | IA | IA | IA | IA | IA | IA | IA | IA | IA | IA | IA | IA | IA | IA | IA | IA | IA | IA | IA | IA | IA | IA | IA | IA | IA | IA | IA | IA | IA | IA | IA | IA |
| WI | WI | WI | WI | WI | WI | WI | WI | WI | WI | WI | WI | WI | WI | WI | WI | WI | WI | WI | WI | WI | WI | WI | WI | WI | WI | WI | WI | WI | WI | WI | WI | WI | WI | WI | WI | WI | WI | WI | WI | WI | WI | WI |
| WY | WY | WY | WY | WY | WY | WY | WY | WY | WY | WY | WY | WY | WY | WY | WY | WY | WY | WY | WY | WY | WY | WY | WY | WY | WY | WY | WY | WY | WY | WY | WY | WY | WY | WY | WY | WY | WY | WY | WY | WY | WY | WY |
| NJ | NJ | NJ | NJ | NJ | NJ | NJ | NJ | NJ | NJ | NJ | NJ | NJ | NJ | NJ | NJ | NJ | NJ | NJ | NJ | NJ | NJ | NJ | NJ | NJ | NJ | NJ | NJ | NJ | NJ | NJ | NJ | NJ | NJ | NJ | NJ | NJ | NJ | NJ | NJ | NJ | NJ | NJ |
| CT | CT | CT | CT | CT | CT | CT | CT | CT | CT | CT | CT | CT | CT | CT | CT | CT | CT | CT | CT | CT | CT | CT | CT | CT | CT | CT | CT | CT | CT | CT | CT | CT | CT | CT | CT | CT | CT | CT | CT | CT | CT | CT |
| NE | NE | NE | NE | NE | NE | NE | NE | NE | NE | NE | NE | NE | NE | NE | NE | NE | NE | NE | NE | NE | NE | NE | NE | NE | NE | NE | NE | NE | NE | NE | NE | NE | NE | NE | NE | NE | NE | NE | NE | NE | NE | NE |
| IN | IN | IN | IN | IN | IN | IN | IN | IN | IN | IN | IN | IN | IN | IN | IN | IN | IN | IN | IN | IN | IN | IN | IN | IN | IN | IN | IN | IN | IN | IN | IN | IN | IN | IN | IN | IN | IN | IN | IN | IN | IN | IN |
| MN | MN | MN | MN | MN | MN | MN | MN | MN | MN | MN | MN | MN | MN | MN | MN | MN | MN | MN | MN | MN | MN | MN | MN | MN | MN | MN | MN | MN | MN | MN | MN | MN | MN | MN | MN | MN | MN | MN | MN | MN | MN | MN |
| VA | VA | VA | VA | VA | VA | VA | VA | VA | VA | VA | VA | VA | VA | VA | VA | VA | VA | VA | VA | VA | VA | VA | VA | VA | VA | VA | VA | VA | VA | VA | VA | VA | VA | VA | VA | VA | VA | VA | VA | VA | VA | VA |
| PA | PA | PA | PA | PA | PA | PA | PA | PA | PA | PA | PA | PA | PA | PA | PA | PA | PA | PA | PA | PA | PA | PA | PA | PA | PA | PA | PA | PA | PA | PA | PA | PA | PA | PA | PA | PA | PA | PA | PA | PA | PA | PA |
| UT | UT | UT | UT | UT | UT | UT | UT | UT | UT | UT | UT | UT | UT | UT | UT | UT | UT | UT | UT | UT | UT | UT | UT | UT | UT | UT | UT | UT | UT | UT | UT | UT | UT | UT | UT | UT | UT | UT | UT | UT | UT | UT |
| OK | OK | OK | OK | OK | OK | OK | OK | OK | OK | OK | OK | OK | OK | OK | OK | OK | OK | OK | OK | OK | OK | OK | OK | OK | OK | OK | OK | OK | OK | OK | OK | OK | OK | OK | OK | OK | OK | OK | OK | OK | OK | OK |
| MO | MO | MO | MO | MO | MO | MO | MO | MO | MO | MO | MO | MO | MO | MO | MO | MO | MO | MO | MO | MO | MO | MO | MO | MO | MO | MO | MO | MO | MO | MO | MO | MO | MO | MO | MO | MO | MO | MO | MO | MO | MO | MO |
| ID | ID | ID | ID | ID | ID | ID | ID | ID | ID | ID | ID | ID | ID | ID | ID | ID | ID | ID | ID | ID | ID | ID | ID | ID | ID | ID | ID | ID | ID | ID | ID | ID | ID | ID | ID | ID | ID | ID | ID | ID | ID | ID |
| OH | OH | OH | OH | OH | OH | OH | OH | OH | OH | OH | OH | OH | OH | OH | OH | OH | OH | OH | OH | OH | OH | OH | OH | OH | OH | OH | OH | OH | OH | OH | OH | OH | OH | OH | OH | OH | OH | OH | OH | OH | OH | OH |
| RI | RI | RI | RI | RI | RI | RI | RI | RI | RI | RI | RI | RI | RI | RI | RI | RI | RI | RI | RI | RI | RI | RI | RI | RI | RI | RI | RI | RI | RI | RI | RI | RI | RI | RI | RI | RI | RI | RI | RI | RI | RI | RI |
| CO | CO | CO | CO | CO | CO | CO | CO | CO | CO | CO | CO | CO | CO | CO | CO | CO | CO | CO | CO | CO | CO | CO | CO | CO | CO | CO | CO | CO | CO | CO | CO | CO | CO | CO | CO | CO | CO | CO | CO | CO | CO | CO |
| MI | MI | MI | MI | MI | MI | MI | MI | MI | MI | MI | MI | MI | MI | MI | MI | MI | MI | MI | MI | MI | MI | MI | MI | MI | MI | MI | MI | MI | MI | MI | MI | MI | MI | MI | MI | MI | MI | MI | MI | MI | MI | MI |
| WV | WV | WV | WV | WV | WV | WV | WV | WV | WV | WV | WV | WV | WV | WV | WV | WV | WV | WV | WV | WV | WV | WV | WV | WV | WV | WV | WV | WV | WV | WV | WV | WV | WV | WV | WV | WV | WV | WV | WV | WV | WV | WV |
| NY | NY | NY | NY | NY | NY | NY | NY | NY | NY | NY | NY | NY | NY | NY | NY | NY | NY | NY | NY | NY | NY | NY | NY | NY | NY | NY | NY | NY | NY | NY | NY | NY | NY | NY | NY | NY | NY | NY | NY | NY | NY | NY |
| DE | DE | DE | DE | DE | DE | DE | DE | DE | DE | DE | DE | DE | DE | DE | DE | DE | DE | DE | DE | DE | DE | DE | DE | DE | DE | DE | DE | DE | DE | DE | DE | DE | DE | DE | DE | DE | DE | DE | DE | DE | DE | DE |
| KY | KY | KY | KY | KY | KY | KY | KY | KY | KY | KY | KY | KY | KY | KY | KY | KY | KY | KY | KY | KY | KY | KY | KY | KY | KY | KY | KY | KY | KY | KY | KY | KY | KY | KY | KY | KY | KY | KY | KY | KY | KY | KY |
| TX | TX | TX | TX | TX | TX | TX | TX | TX | TX | TX | TX | TX | TX | TX | TX | TX | TX | TX | TX | TX | TX | TX | TX | TX | TX | TX | TX | TX | TX | TX | TX | TX | TX | TX | TX | TX | TX | TX | TX | TX | TX | TX |
| GA | GA | GA | GA | GA | GA | GA | GA | GA | GA | GA | GA | GA | GA | GA | GA | GA | GA | GA | GA | GA | GA | GA | GA | GA | GA | GA | GA | GA | GA | GA | GA | GA | GA | GA | GA | GA | GA | GA | GA | GA | GA | GA |
| TN | TN | TN | TN | TN | TN | TN | TN | TN | TN | TN | TN | TN | TN | TN | TN | TN | TN | TN | TN | TN | TN | TN | TN | TN | TN | TN | TN | TN | TN | TN | TN | TN | TN | TN | TN | TN | TN | TN | TN | TN | TN | TN |
| NC | NC | NC | NC | NC | NC | NC | NC | NC | NC | NC | NC | NC | NC | NC | NC | NC | NC | NC | NC | NC | NC | NC | NC | NC | NC | NC | NC | NC | NC | NC | NC | NC | NC | NC | NC | NC | NC | NC | NC | NC | NC | NC |
| MD | MD | MD | MD | MD | MD | MD | MD | MD | MD | MD | MD | MD | MD | MD | MD | MD | MD | MD | MD | MD | MD | MD | MD | MD | MD | MD | MD | MD | MD | MD | MD | MD | MD | MD | MD | MD | MD | MD | MD | MD | MD | MD |
| AR | AR | AR | AR | AR | AR | AR | AR | AR | AR | AR | AR | AR | AR | AR | AR | AR | AR | AR | AR | AR | AR | AR | AR | AR | AR | AR | AR | AR | AR | AR | AR | AR | AR | AR | AR | AR | AR | AR | AR | AR | AR | AR |
| NM | NM | NM | NM | NM | NM | NM | NM | NM | NM | NM | NM | NM | NM | NM | NM | NM | NM | NM | NM | NM | NM | NM | NM | NM | NM | NM | NM | NM | NM | NM | NM | NM | NM | NM | NM | NM | NM | NM | NM | NM | NM | NM |
| SC | SC | SC | SC | SC | SC | SC | SC | SC | SC | SC | SC | SC | SC | SC | SC | SC | SC | SC | SC | SC | SC | SC | SC | SC | SC | SC | SC | SC | SC | SC | SC | SC | SC | SC | SC | SC | SC | SC | SC | SC | SC | SC |
| AZ | AZ | AZ | AZ | AZ | AZ | AZ | AZ | AZ | AZ | AZ | AZ | AZ | AZ | AZ | AZ | AZ | AZ | AZ | AZ | AZ | AZ | AZ | AZ | AZ | AZ | AZ | AZ | AZ | AZ | AZ | AZ | AZ | AZ | AZ | AZ | AZ | AZ | AZ | AZ | AZ | AZ | AZ |
| FL | FL | FL | FL | FL | FL | FL | FL | FL | FL | FL | FL | FL | FL | FL | FL | FL | FL | FL | FL | FL | FL | FL | FL | FL | FL | FL | FL | FL | FL | FL | FL | FL | FL | FL | FL | FL | FL | FL | FL | FL | FL | FL |
| AL | AL | AL | AL | AL | AL | AL | AL | AL | AL | AL | AL | AL | AL | AL | AL | AL | AL | AL | AL | AL | AL | AL | AL | AL | AL | AL | AL | AL | AL | AL | AL | AL | AL | AL | AL | AL | AL | AL | AL | AL | AL | AL |
| LA | LA | LA | LA | LA | LA | LA | LA | LA | LA | LA | LA | LA | LA | LA | LA | LA | LA | LA | LA | LA | LA | LA | LA | LA | LA | LA | LA | LA | LA | LA | LA | LA | LA | LA | LA | LA | LA | LA | LA | LA | LA | LA |
| HI | HI | HI | HI | HI | HI | HI | HI | HI | HI | HI | HI | HI | HI | HI | HI | HI | HI | HI | HI | HI | HI | HI | HI | HI | HI | HI | HI | HI | HI | HI | HI | HI | HI | HI | HI | HI | HI | HI | HI | HI | HI | HI |
| CA | CA | CA | CA | CA | CA | CA | CA | CA | CA | CA | CA | CA | CA | CA | CA | CA | CA | CA | CA | CA | CA | CA | CA | CA | CA | CA | CA | CA | CA | CA | CA | CA | CA | CA | CA | CA | CA | CA | CA | CA | CA | CA |
| MS | MS | MS | MS | MS | MS | MS | MS | MS | MS | MS | MS | MS | MS | MS | MS | MS | MS | MS | MS | MS | MS | MS | MS | MS | MS | MS | MS | MS | MS | MS | MS | MS | MS | MS | MS | MS | MS | MS | MS | MS | MS | MS |
| DC | DC | DC | DC | DC | DC | DC | DC | DC | DC | DC | DC | DC | DC | DC | DC | DC | DC | DC | DC | DC | DC | DC | DC | DC | DC | DC | DC | DC | DC | DC | DC | DC | DC | DC | DC | DC | DC | DC | DC | DC | DC | DC |
| GU | GU | GU | GU | GU | GU | GU | GU | GU | GU | GU | GU | GU | GU | GU | GU | GU | GU | GU | GU | GU | GU | GU | GU | GU | GU | GU | GU | GU | GU | GU | GU | GU | GU | GU | GU | GU | GU | GU | GU | GU | GU | GU |

State has statistically significantly higher average proficiency than the state listed at the top of the chart.

No statistically significant difference from the state listed at the top of the chart.

State has statistically significantly lower average proficiency than the state listed at the top of the chart.

The between state comparisons take into account sampling and measurement error and that each state is being compared with every other state. Significance is determined by an application of the Bonferroni procedure.

*Did not statisfy one or more of the guidelines for sample participation rates (see Appendix for details).

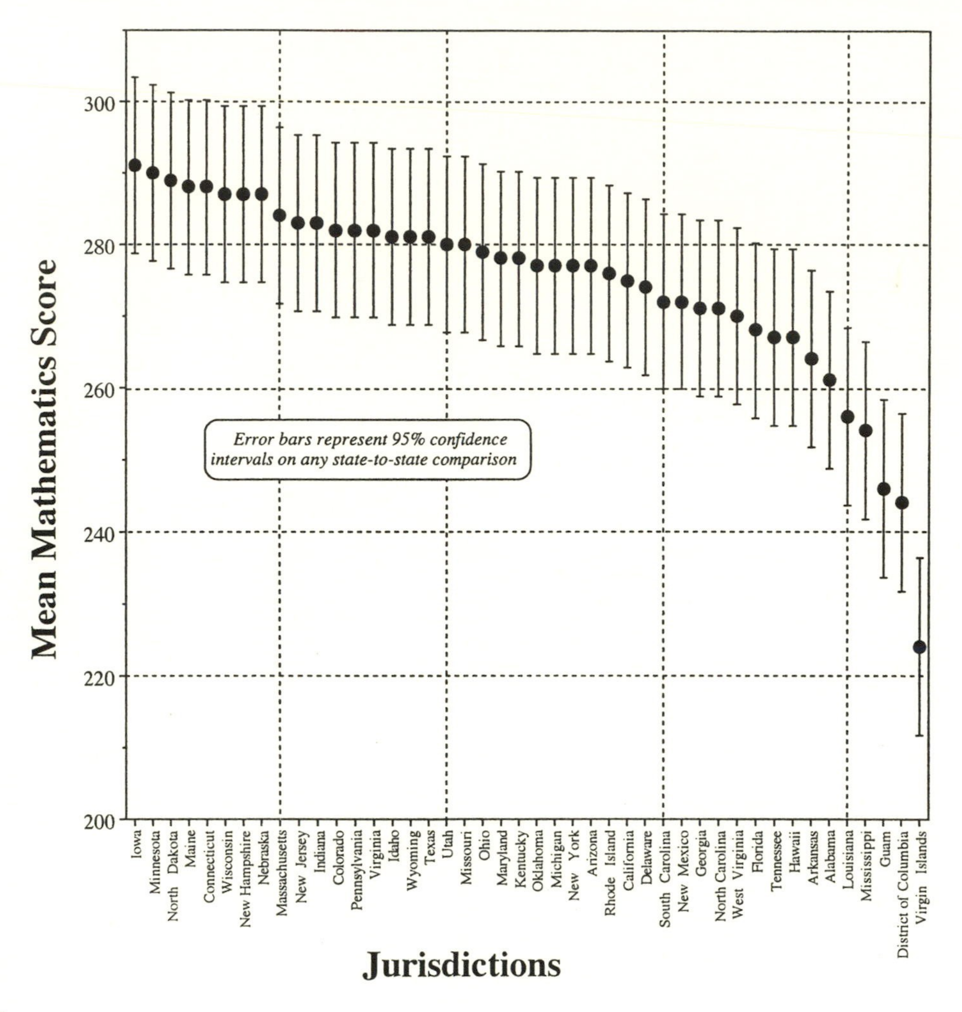


**Figure 13.7.**
Average mathematics proficiency for eighth grade children with at least one parent who graduated from college. 1992 NAEP State Assessment data. A difference between two states is statistically significant (at the .05 level) if one state's data point is outside the other state's bounds.

is more error-laden. Thus we might draw less precisely estimated points with a large symbol (a longer bar) and more precisely estimated ones with a smaller symbol (a shorter bar). Then when fitting we try to make the fitted function pass through all the error bars. Often, however, we can implement this sort of weighted regression analytically and do not have to clog the display visually with inaccurate points.

A basic premise of effective display[5] is to make the visual representation of the data an accurate visual metaphor. If the data increase the

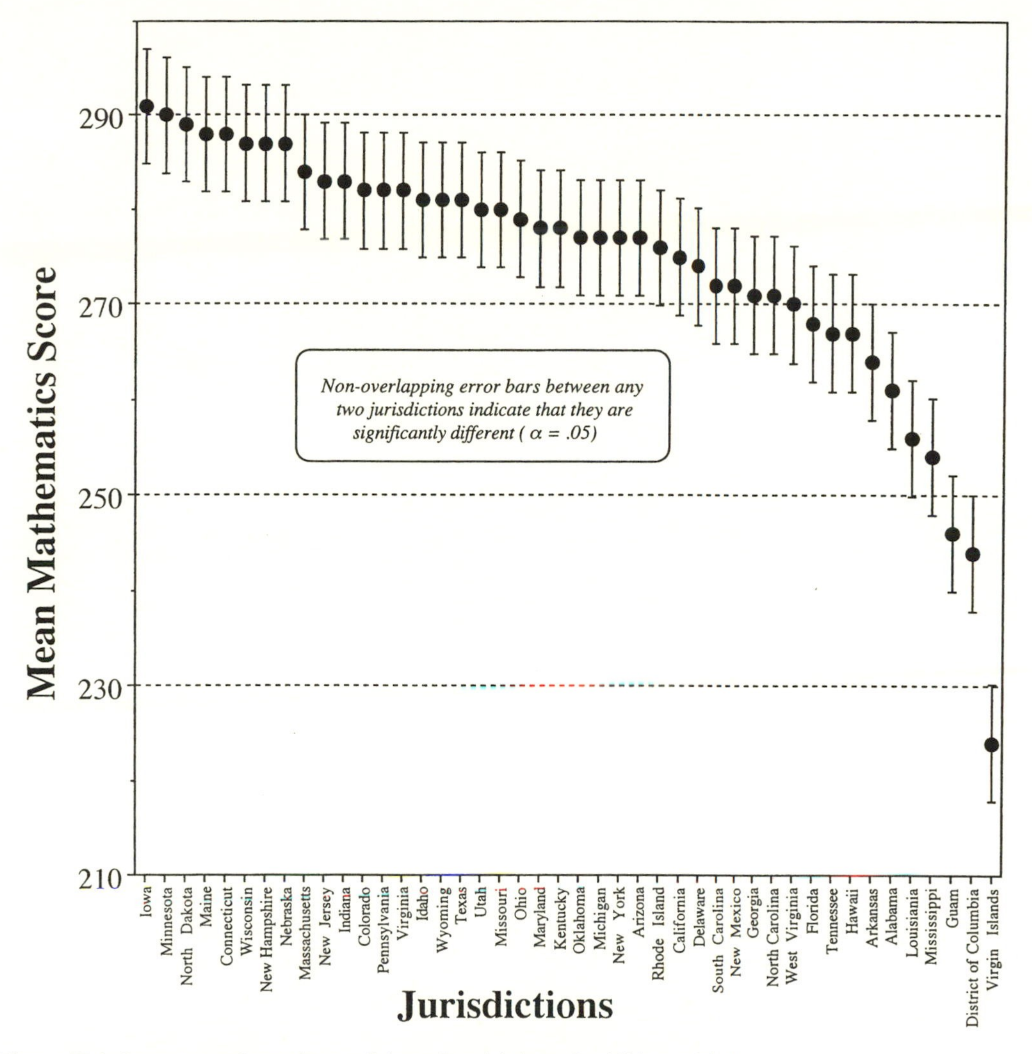


**Figure 13.8.** Average mathematics proficiency for eighth grade children with at least one parent who graduated from college. 1992 NAEP State Assessment data. A difference between two states is statistically significant (at the .05 level) if one state's bounds overlap the other state's bounds.

representation should get larger or go higher; more should indicate more. A display that violates this rule is bound to be misunderstood. How can this rule be implemented in depicting data and their associated error? At first blush it would seem to suggest that bigger error

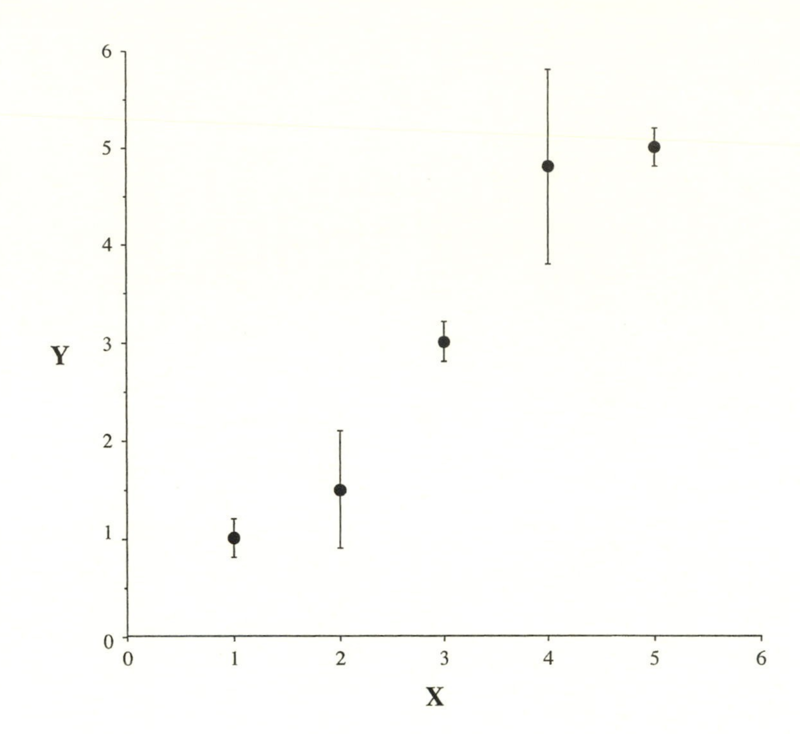


Figure 13.9.
A plot of hypothetical data that displays the data and their standard errors as vertical bars. The graph suggests a nonlinear relationship between *X* and *Y*.

bars should be associated with bigger error. Indeed this is probably correct if the focus of the display is on the error. If, however, the focus is not on the error but rather on the data themselves, perhaps we can improve the comprehensibility of the display by reversing this metaphor; by making the size of the plotting icon proportional to the precision and not the error. Thus data points that are more accurately estimated are drawn to make a bigger visual impression. This suggests that we make the data points' size proportional to their precision (i.e., perhaps proportional to 1/standard error if we wish to focus on the precision). In this way our attention will be drawn to the more accurate points; very inaccurate points will be hardly visible and so have little impact.

As an example consider the hypothetical data shown in figure 13.9 There are five points shown with their standard errors. The graph suggests that the relationship between *X* and *Y* is nonlinear, with an S-shape. The same data are shown in figure 13.10, in which the data points' areas are shown proportional to their precision (you must look

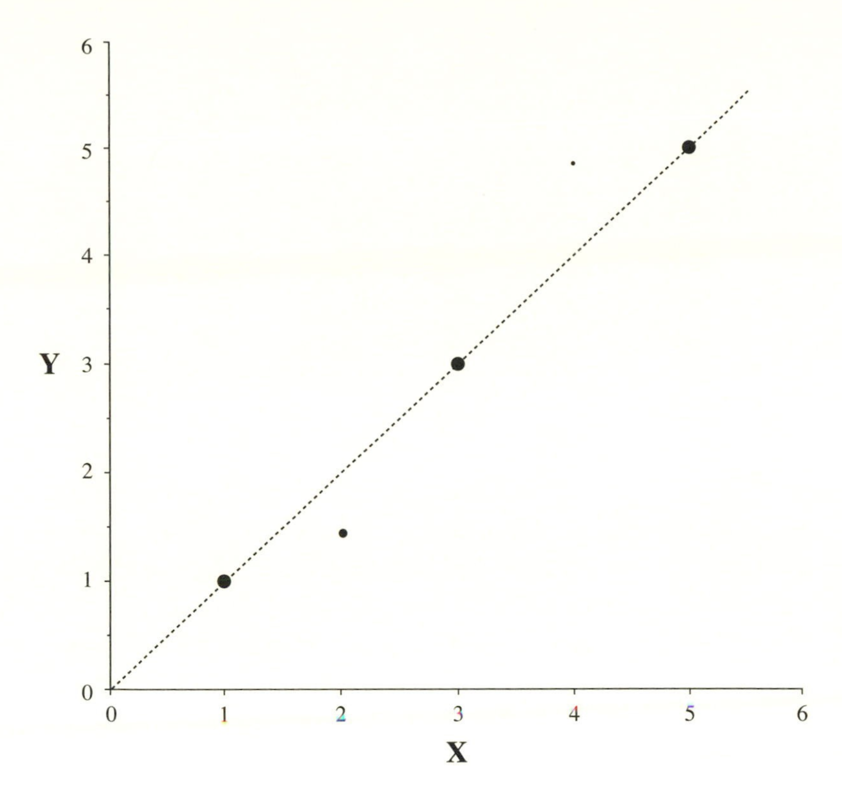


**Figure 13.10.**
A plot of hypothetical data that displays the data sized proportionally to their precision. The graph suggests a linear relationship between $X$ and $Y$.

carefully for point 4). In this depiction the linearity of the data is one's principal perception; such a perception is supported by formal statistical testing. Perhaps we can generalize this idea through the use of something akin to horizontal "precision bars" around each point. These bars could be proportional to the information contained in each point.[6] Another alternative might be to use a plotting icon like those shown below[7] to reflect precision.

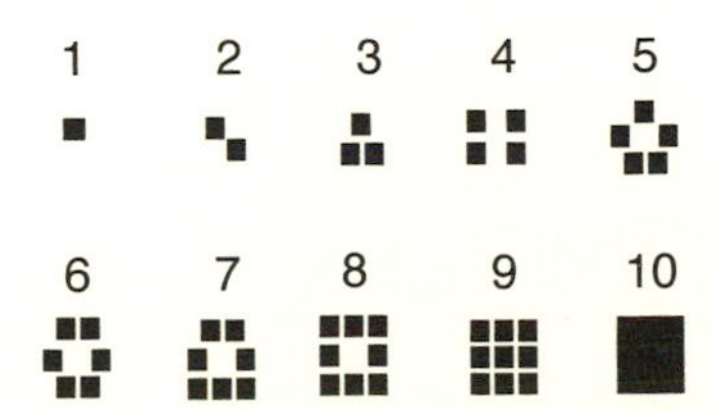

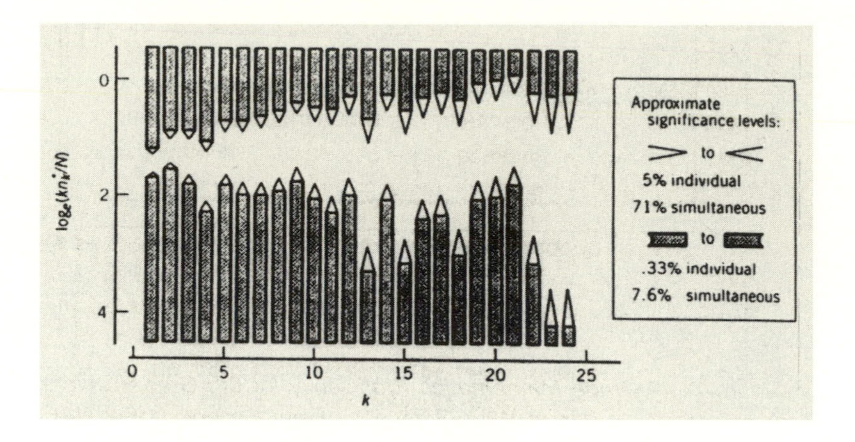


**Figure 13.11.**
A confidence-aperture plot from Hoaglin and Tukey (1985, figure 9-14, p. 386) in which the amount of open space reflects the uncertainty. Reprinted with permission.

David Hoaglin and John Tukey[8] used the amount of empty space to depict the least accurately estimated points in their confidence-aperture plots. Figure 13.11 shows that there is no significant departure from a logarithmic series distribution in the butterfly data that were collected by some earlier investigators.[9] Hoaglin and Tukey plotted a function of the number of species found ($n_k$) and the number of individual members of a butterfly species caught ($k$), against $k$. In this display the tips of the "pencils" represent a 95% confidence interval for an individual data point, while simultaneously yielding but a 29% simultaneous interval for the twenty-four values of $k$. The portion of the plot associated with the top of the body of the pencil yields a larger area and so the corresponding probabilities grow to 99.6% and 92.4%. Hoaglin, Mosteller, and Tukey thought enough of this idea to feature it on the cover of their book.

The worth of such schemes will require some visual experimentation with formats, some experience to overcome the older conventions, and the careful consideration of the questions that the display is meant to answer.

### 13.3.2. Sloppy (Dirty) Graphics

Another graphical way to remind the viewer of the fallibility of the data in a display is to draw and label the display only as well as the accuracy of the data allow. This is exactly analogous to the too often ignored practice of writing down numerical values to only the number of digits that are significant.[10]

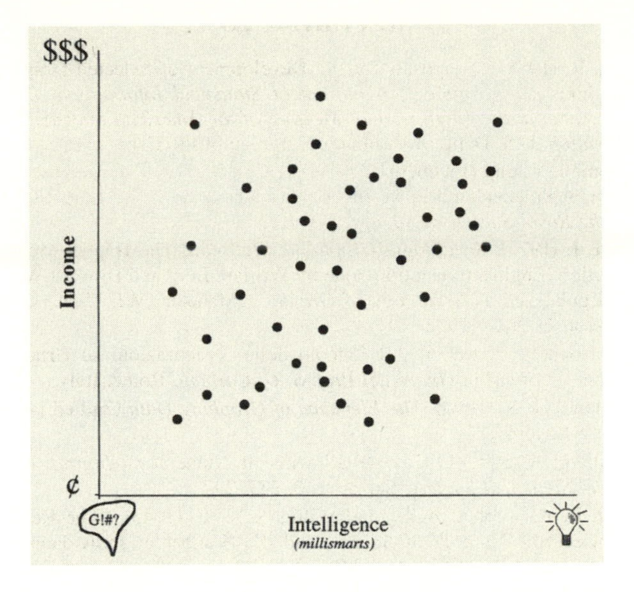


Figure 13.12.
Displaying data of uncertain precision with the accuracy they deserve through the use of appropriately calibrated axes.

One possible approach might be to calibrate and label the axes in such a way as to limit the accuracy of the inferences to only what is justified by the data. An example of such a display is shown in figure 13.12.[11]

While an asset of such a display is that it doesn't allow incorrectly precise statements, a possible drawback is that it doesn't allow imprecisely correct statements. This provokes the image of Damon Runyun, when he said, "The race is not always to the swift, nor the battle to the strong, but that's the way to lay your bets."* Thus, even though we may not have good estimates, it may sometimes be important for the graph to allow us to extract the best guess available.†

An allied approach might be to accurately label and calibrate the axes but remove distinct boundaries from the graphical representation. For example, in figure 13.13 we show the performance of five ethnic

*Perhaps, specializing John Tukey's well known aphorism, this corresponds to "If it's worth displaying, it's worth displaying badly."

†I have often witnessed requests for a single number to characterize a complex situation. Some then argue that "a single number is inappropriate and we ought not provide it." Others cynically contend that "if we don't give them a number they'll calculate one of their own. Let's at least provide the best single number we can." I have some sympathy with both views. Perhaps by providing graphical answers like that in figure 13.12 we can provide the qualitative answer without allowing inferences of inappropriate precision.

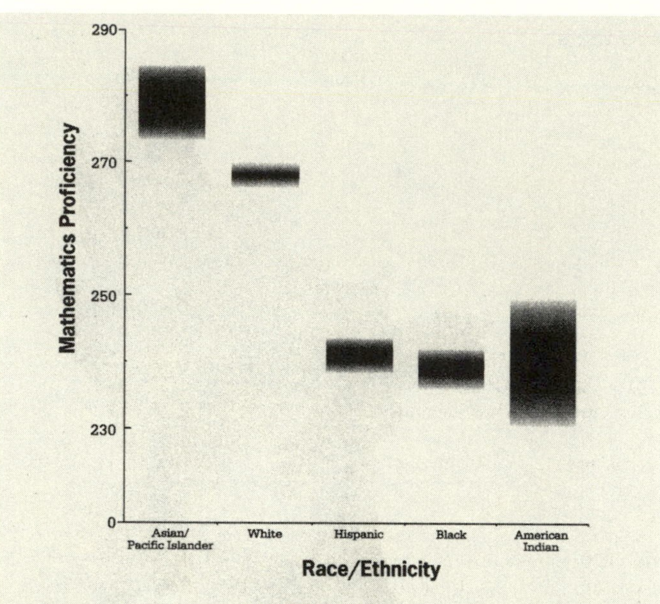


*Figure 13. Average U.S. Proficiency in "Algebra and Functions," Grade 8—-1990. Using graduated shading can make it possible to automatically build the perception of variability into what used to be a bar chart.*

**Figure 13.13.**
Using graduated shading can make it possible to automatically build the perception of variability into what used to be a bar chart.

groups on one portion of the 1992 NAEP mathematics assessment. Through the utilization of graduated shading around the actual mean performance we make it impossible to unambiguously choose a single number to characterize that ethnic group's performance. The location at which the shading begins to thin is determined by both the mean level and the estimated standard error. Groups like American Indians with large standard errors have the shading begin well below the mean and it changes very gradually. Ethnic groups like Whites that have small standard errors are characterized by a much steeper gradient in the shading algorithm.* This approach is a continuous analog† to the more traditional methods shown in figure 13.4.

*Figure 13.13 is little more than an initial attempt. I fully expect that it can be improved upon with experience. The rule used to generate it was to use black at the mean and then graduate the shading exponentially to white 2.5 standard errors above and below the mean.

†This is probably more properly called a Bayesian approach since it depicts a posterior distribution rather than a confidence region.

**Figure 13.14.** Graduated shading was used by Quetelet in 1831 to show the distribution crimes against property (left side) and crimes against persons (right side) (from Robinson, 1982).

## 13.4. SUMMARY

In this chapter I have described and illustrated some suggestions for alternative methods to depict error. The principal goal was to make it easier for inferences about data to include the effects of the precision with which those data were measured. Sometimes (with tabular presentation) this involved calculating and displaying the appropriate error term for the most likely uses of the data. In other situations (like the fade-away chart in figure 13.13) it meant making the precision an integral part of the display. I'm sure that the notion of using shading to characterize error is a long way from being original with me (my former colleague, Albert Biderman, suggested it to me over coffee three decades ago). I am certain that the primary reason we have not seen it used more widely in practice is its difficult implementation. Modern computer graphic software has removed this impediment. This apparently novel statistical suggestion is old hat to cartographers. Petermann (1851) used continuous tone implemented by lithographic crayon shading to show variations in the population of Scotland. But even this use was presaged by Quetelet (1832) who produced some fuzzy social data maps showing the distribution of crime (see figure 13.14). More recently, MacEachren[12] used this method effectively to display the distribution of risk surrounding a nuclear power plant. He also suggests schemes such as defocusing images to convey uncertainty.

Dynamic possibilities for the representation of uncertainty have not been discussed, although we believe that such schemes as blinking data points in which the proportion of "on" to "off" time is proportional to the points' precision, or a visual analog of multiple imputation in which the location of a data point changes as you watch it, are promising possibilities.

The suggestions contained here are meant only as a beginning. The assessment of their usefulness awaits broader application, discussion, and experimentation for, reapplying Karl Pearson's final words in the *Grammar of Science*, it is only by "daring to display our ignorance" that our perceptions of scientific progress will stay in reasonably close proximity to that progress.

## 13.5. A BRIEF TUTORIAL ON STATISTICAL ERROR

Chapter 13 is, in some sense, the most technical one in this book. The ideas are not difficult, but the terminology may be unfamiliar. So let me take a few pages to provide some background information and definitions. This section may be skipped by those readers who remember this material from an introductory statistics course.

First, *hypothesis testing*. Although the basic ideas of hypothesis testing were laid out in 1710 by the physician and satirist John Arbuthnot, the modern formal theory is usually credited to a collaboration between the Polish mathematician Jerzy Neyman and the British statistician Egon Pearson. The idea is that we specify competing hypotheses usually called $H_0$, the null hypothesis, and $H_1$, the alternative. These two hypotheses must be mutually exclusive and exhaustive.

As a concrete example, let us consider an experiment in which we try out a new drug treatment for some horrible disease. We randomly assign half the people with the disease to the existing treatment and the other half to the new drug. Then we gather evidence about the efficacy of the two treatments, specifically how long it takes to be cured. We calculate the average time to cure under each of the conditions as well as how much variability there was in each group. So the average time to cure for the new treatment might be 100 days, but some people might be cured in 50 days and others in 150. The average time to cure for the old treatment might be 110 days, with similar variability. We recognize that if we were to repeat the experiment with different people the results would not be exactly the same. So we conceptualize the question as follows:

> The average that we calculated for each of the two groups was only an estimate of what the true average would have been had we tested the entire population of people with the illness. We ask "was the difference between the two treatments large enough for us to decide that the population values were not likely to be the same?"

Formally, this is done by calling the true mean cure time for the new treatment $\mu_T$ and the true mean for the old treatment (now the control treatment) $\mu_C$, and the means that we actually observed are merely estimates from a finite sample. This notion of a sample being an imperfect reflection of the population value is a direct lineal descendent of the shadows on the wall of Plato's cave.

We then set up the competing hypotheses as

$$H_0: \mu_T = \mu_C.$$

and

$$H_1: \mu_T \neq \mu_C.$$

Note that they are exhaustive, there are no other alternatives to these two, and they are mutually exclusive.

The evidence that we gather is never going to yield certainty, since we will never be able to treat everyone. There are two kinds of errors that we can make. We can reject $H_0$ when it is true—this is called a *Type I error*. Or we do not reject it when it is false—this is a *Type II error*. These two kinds of errors yield a natural and diabolical tradeoff. If we try to make the Type I error as small as possible, we must increase the Type II error. This is obvious with only a little thought, for if we are so cautious that we never reject $H_0$ we have reduced the Type I error to zero, but have maximized the Type II error. So the goal of good experimentation is to control Type I while minimizing Type II.

The probability of a Type I error is generally denoted with the Greek letter $\alpha$, and a nominal value that is often used for preliminary experimentation is .05, or one chance in twenty.

What this means in our *gedanken* drug experiment is that when the difference between the experimental and control treatments reaches a specific level we will reject the null hypothesis and when we do so, we will be wrong about one time in twenty. How big the difference must be between the two observed means to achieve a significant difference depends on how much variability there is among the experimental subjects. If they are very variable, the difference must be larger because the variability observed within each group tells us that the mean computed is not very stable (remember De Moivre's equation discussed in chapter 1). And most of chapter 13 is devoted to displaying how much variability there is in comparison to the mean level.

So much for the basic idea of hypothesis testing; next let us consider *the problem of multiplicity*. Suppose we do more than a single experiment. So instead of testing just a single experimental treatment, we test two of them. The chance of our not rejecting the null hypothesis incorrectly is 0.95 (or $\alpha$) on either one of the tests, and thus being correct on both of them is $0.95 \times 0.95$ or 0.90. So, the probability of a Type I error for the set of two experiments is no longer the one in twenty ($\alpha = .05$) that the experiment was designed for, but rather 0.10. And if we were testing three different drugs it would be $1 - (0.95)^3 = 1 - 0.86 = 0.14$.

Thus the original controls on Type I errors that we built into the experiment are diluted as the size of the family of tests increases. What can we do to fix this? Happily, there are many methods that have been proposed. The one I will focus on here is the simplest and yet is both general and efficient. It was developed by the Italian mathematician Carlo Emilio Bonferroni (1892–1960) and is named the *Bonferroni method* in his honor. Simply stated, if we are doing $n$ tests we simply divide the $\alpha$ level of the family of tests by $n$, and to be considered a significant difference any individual tests must be significant at that level. So if we do two tests and we want the $\alpha$ of the family of tests to be 0.05, to be considered significant at this level an individual test must be significant at $0.05/2 = 0.025$, or one in forty. If we do three tests each must be significant at $0.05/3 = 0.017$ and so on. Easy!

Doing a correction for multiplicity is very important. If we didn't and we performed twenty tests at the 0.05 level we would expect one of them to show up as significant just by chance. By doing a formal correction we keep control of error rates on what might otherwise be just some sort of exploratory fishing expedition.

You can see why the Bonferroni correction is so popular. It is easy to do and works very well. But there are some circumstances where the demands that it makes on data are so stringent that, if it was the only method used, it would essentially preclude ever finding anything. For example, in a study of the genetic basis of glaucoma among Mennonites (an unusually frequent occurrence among them) scientists found one hundred Mennonites with glaucoma and matched them on many characteristics to one hundred other Mennonites without glaucoma. They then compared the two groups on the frequency of occurrence of a particular component on each of over 100,000 genes. Initially they planned to do 100,000 *t*-tests. If they wanted an overall $\alpha$ level of 0.05, a Bonferroni adjustment would mean that to be considered significant any particular gene must be different between the two groups at $0.05/100{,}000 = 0.00000005$. This is a level of difference unlikely to be possible within the practical confines of the experiment. For such circumstances other methods for controlling multiplicity must be employed.

Let us leave the realm of hypothesis testing now for some brief definitions of other terms used in the chapter. Specifically,

*Sensitivity analysis* is a procedure that examines the extent to which the conclusions we draw are sensitive to minor variations in the data. If the conclusions are robust with respect to smallish bits of variability, confidence in the outcome is increased. If even tiny variations cause large changes in our conclusions we have been warned.

*Resampling* is one way of doing such sensitivity studies. Instead of using the entire data sample to draw conclusions we instead sample from it, with replacement, build a new sample, and see if the same conclusions hold in this sample. Then redo it many times. As an example suppose the original data set has 100 observations, we can build many subsamples from it by taking one observation out of it at random and recording it and then putting it back in (this is what "with replacement" means) and then taking another. Continue until, say, 90 or even 100 are taken. Of course, since we are replacing each observation after we record it, the same observation might recur more than once in our subsample. The we redo this over and over again. Each subsample (this was dubbed a "bootstrap" sample by Stanford statistician Brad Efron, who invented it, since it appears that we are lifting ourselves up by our own bootstraps) will have the same general characteristics as the original, but will exhibit some variability. We can use this variability to see how much variation in our conclusions is elicited by this internal variation.

*Multiple imputation* is another method. It is usually used when there are missing data points and we wish to know how much our conclusions are influenced by what we did not observe. So first we draw whatever conclusions we think justified by the data observed. Then we insert (impute) values where there were none and see how our conclusions change. Then impute different numbers and see what we conclude. And continue doing this until the numbers we have imputed span the entire range of what is plausible. The results of this provide us with a measure of how sensitive our results are to what was missing. As an example, suppose we wish to estimate the average number of occupants in each apartment of a building. We might find that of the one hundred apartments in the building fifty of them have one occupant and forty have two. Of the remaining ten, no one answered when we rang their doorbell. We might impute one person for those ten and then calculate that the mean number of occupants is $(50 + 10 + 80)/100 = 1.4$. Or we could impute two people in each apartment and calculate $(50 + 20 + 80)/100 = 1.5$. Or we could insert zero and get 1.3. Or impute three and get 1.6. Or anything in between . After doing this we could say that the mean number of occupants per apartment probably is between 1.3 and 1.6. Of course our imputations could be wrong, and the ten apartments didn't answer our ring because they were hiding ten large families of illegal immigrants, but without additional information that is sufficiently unlikely to prevent us from being satisfied with the error bounds obtained. Note that, as the amount of missing data increases, so too do the error bounds. Consider what the error bounds of a survey are when there is only a 25% response rate. Can we have any faith in the results? Only if we assume that the people who did not respond are just like the ones who did (called "missing at random"), but that is rarely plausible, since by choosing not to respond they indicated that they are, in some ways, not the same.

Last, I used the term *standard error* more often than seldom. It does not mean the error that most people make. Rather it has a very specific, technical meaning. In few words, it is the positive square root of the variance of the sampling distribution of a statistic. In more words, and as an example of what this means, suppose we have a statistic of interest, say the sample mean. If we calculate the sample mean many times, from many different samples of the same size, these means (what we call the sampling distribution of the mean) will not all be the same; they will demonstrate some variability. We could measure this variability by calculating the variance of this sampling distribution. If we then take the positive square root of this variance it is the standard error, and it characterizes how much we expect the sample mean to bounce around. Thus, going back to chapter 1, De Moivre's equation is a way of calculating the standard error of the mean from the population standard deviation.

# 14

## The Mendel Effect

### 14.1. INTRODUCTION

Anyone who listens to baseball (or most other kinds of sporting events) has heard the announcer proclaim a statement akin to "Fred is seven for his last twelve at bats." The implication is that Fred is a heck of a hitter and is on a tear. Implicit in this statement is that Fred is also seven for at least his last thirteen at bats, and might be seven for his last thirty or forty. The announcer chose 12 as his window of calculation because it best supported his thesis that Fred is on a hot streak. It also, however, provides a biased estimate of Fred's ability.

Such biased reporting is not limited to sports announcers. Johann Gregor Mendel (1822–1884) is generally credited with being the father of modern genetics. His research, between 1856 and 1863, focused on the hybridization of peas. He posited that smooth and wrinkled skins of peas were genetic and could be represented by a dominant and a recessive allele, say A and *a*. Thus the possible genotypes were AA, A*a*, and *aa*, but only the doubly recessive seeds *aa* have a different phenotype (appearance). Further, if both alleles are equally represented in the population we should expect 25% of the genotypes to be *AA*, 50% A*a*, and 25% *aa*; however, the two distinguishable phenotypes should be distributed 75%:25%. To test his theory he ran experiments and kept meticulous records. He found that the data he gathered supported his

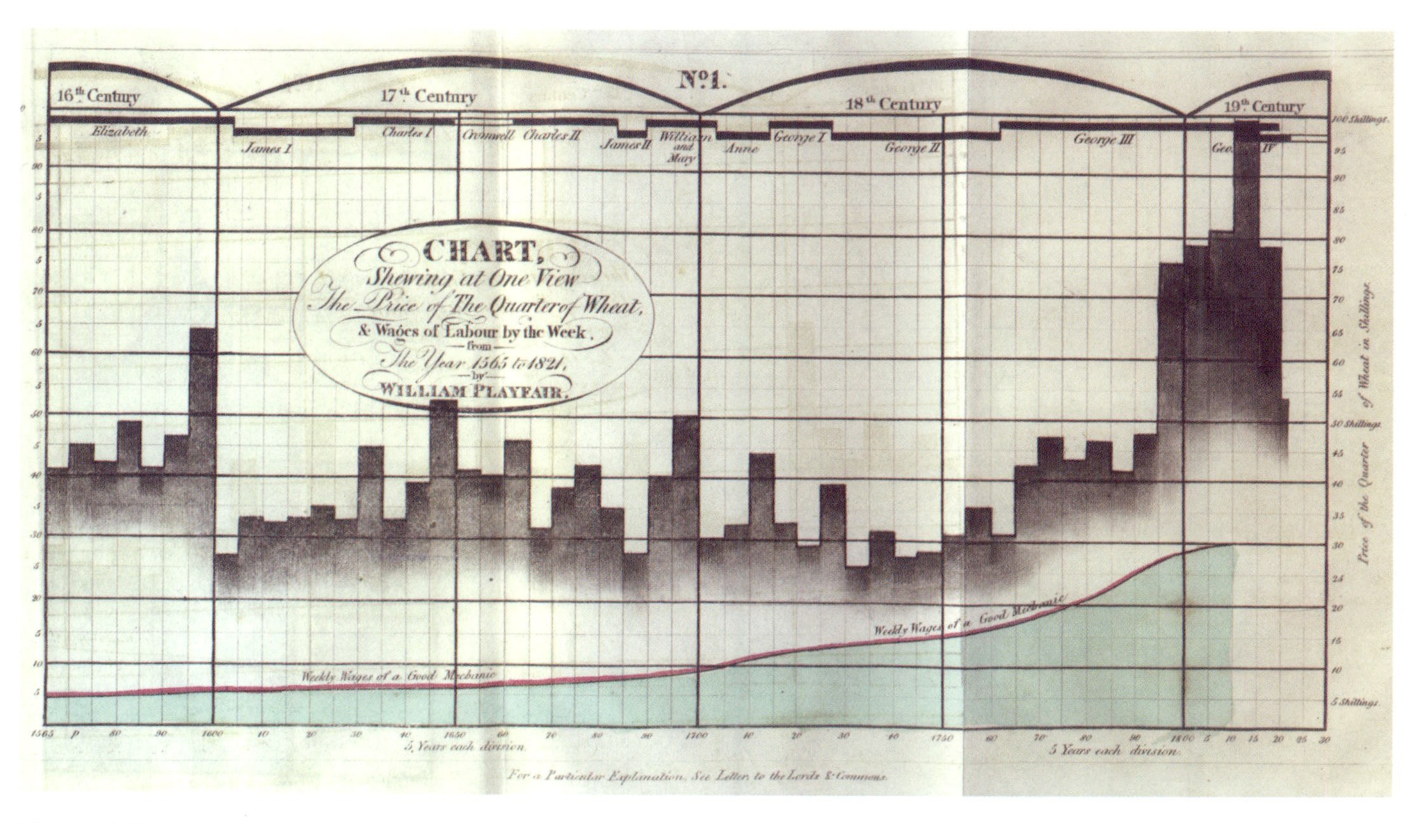


**Figures 11.12 and 17.3.** A time series display showing three parallel time series: prices of a quarter of wheat (the histogram bars), wages of a good mechanic (the line beneath it), and the reigns of English monarchs from Elizabeth I to George IV (1565–1820). Our gratitude for this copy of Playfair's figure to Stephen Ferguson and the Department of Rare Books and Special Collections, Princeton University Library.

*http://www.Florence-Nightingale-Avenging-Angel.co.uk/Coxcomb.htm*

DIAGRAM OF THE CA

IN THE ARMY

2.

APRIL 1855 TO MARCH 1856.

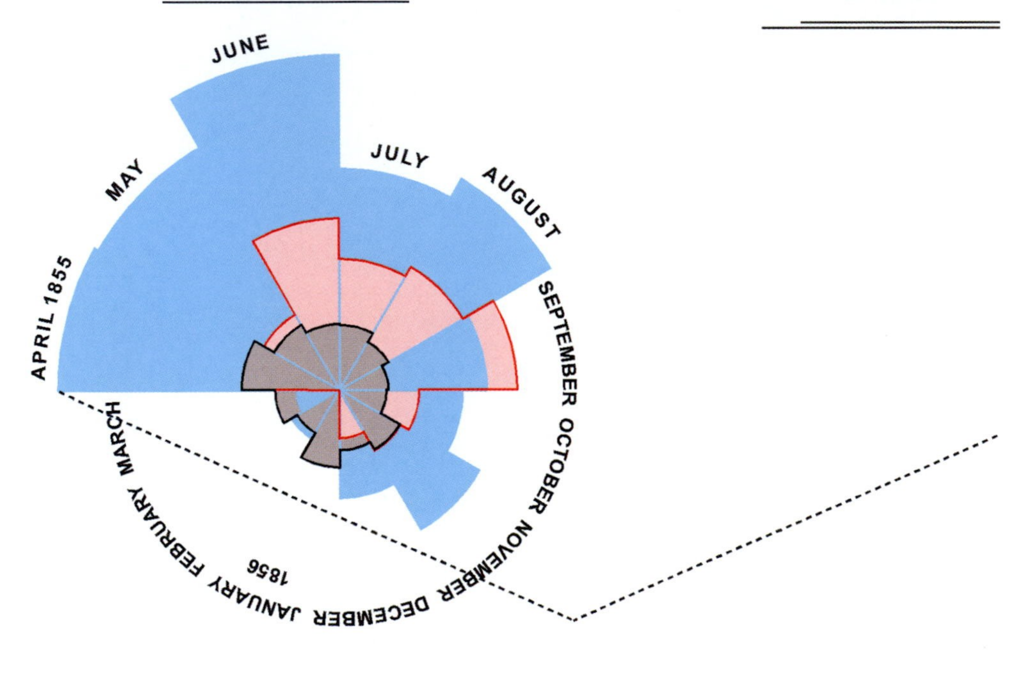


*The Areas of the blue, red, & black wedges are each measured from the centre as the common vertex*

*The blue wedges measured from the centre of the circle represent area for area the deaths from Preventible or Mitigable Zymotic Diseases, the red wedges measured from the centre the deaths from wounds, & the black wedges measured from the centre the deaths from all other causes*

*The black lines across the red triangles in Sept & Nov 1854 mark the boundaries of the deaths from all other causes during those months*

*In October 1854, April 1855, & November 1855, the black area coincides with the red, in January & February 1856, the blue coincides with the black*

*The entire areas may be compared by following the blue, the red & the black lines enclosing them*

**Figure 11.5.** A redrafting of Florence Nightingale's famous "coxcomb" display (what has since become known as a Nightingale Rose) showing the variation in mortality over the months of the year.

*Diagram by Florence Nightingale, corrected by Hugh Small*

USES OF MORTALITY
IN THE EAST.

1.
APRIL 1854 TO MARCH 1855.

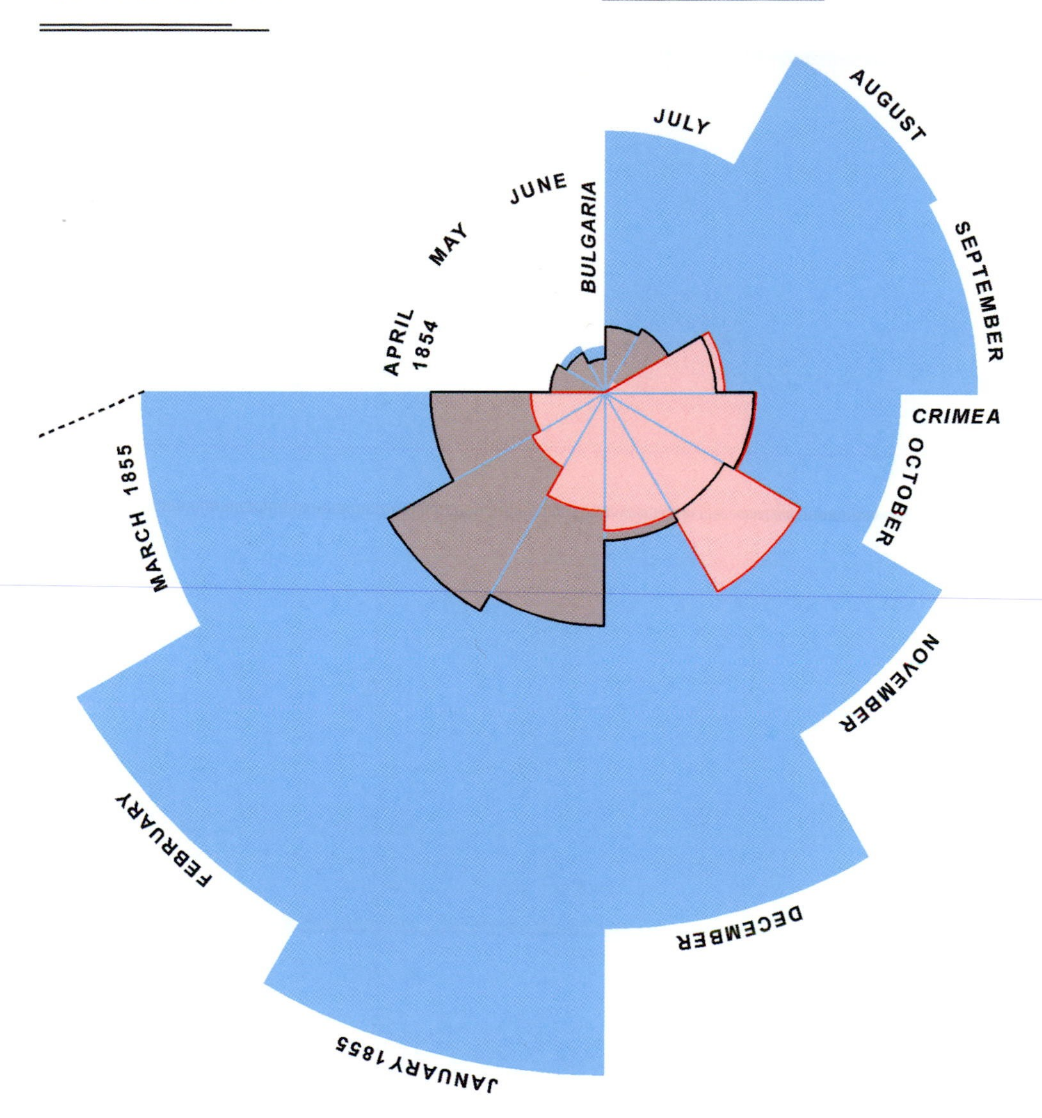


Figure 11.5. (*Continued*)

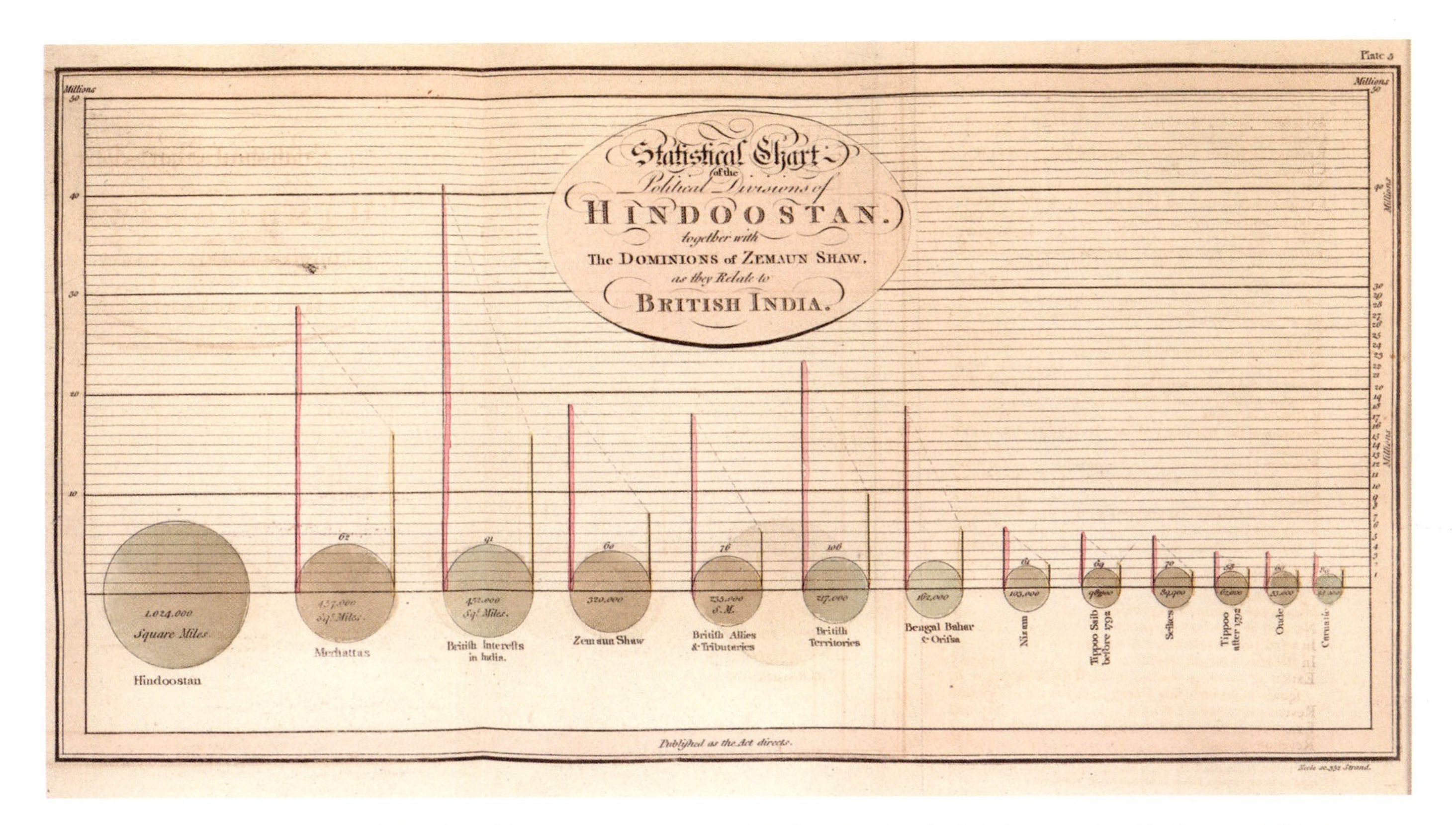


Figure 11.13. A graph taken from Playfair (1801). It contains three data series. The area of each circle is proportional to the area of the geographic location indicated. The vertical line to the left of each circle expresses the number of inhabitants, in millions. The vertical line to the right, represents the revenue generated in that region in millions of pounds sterling.

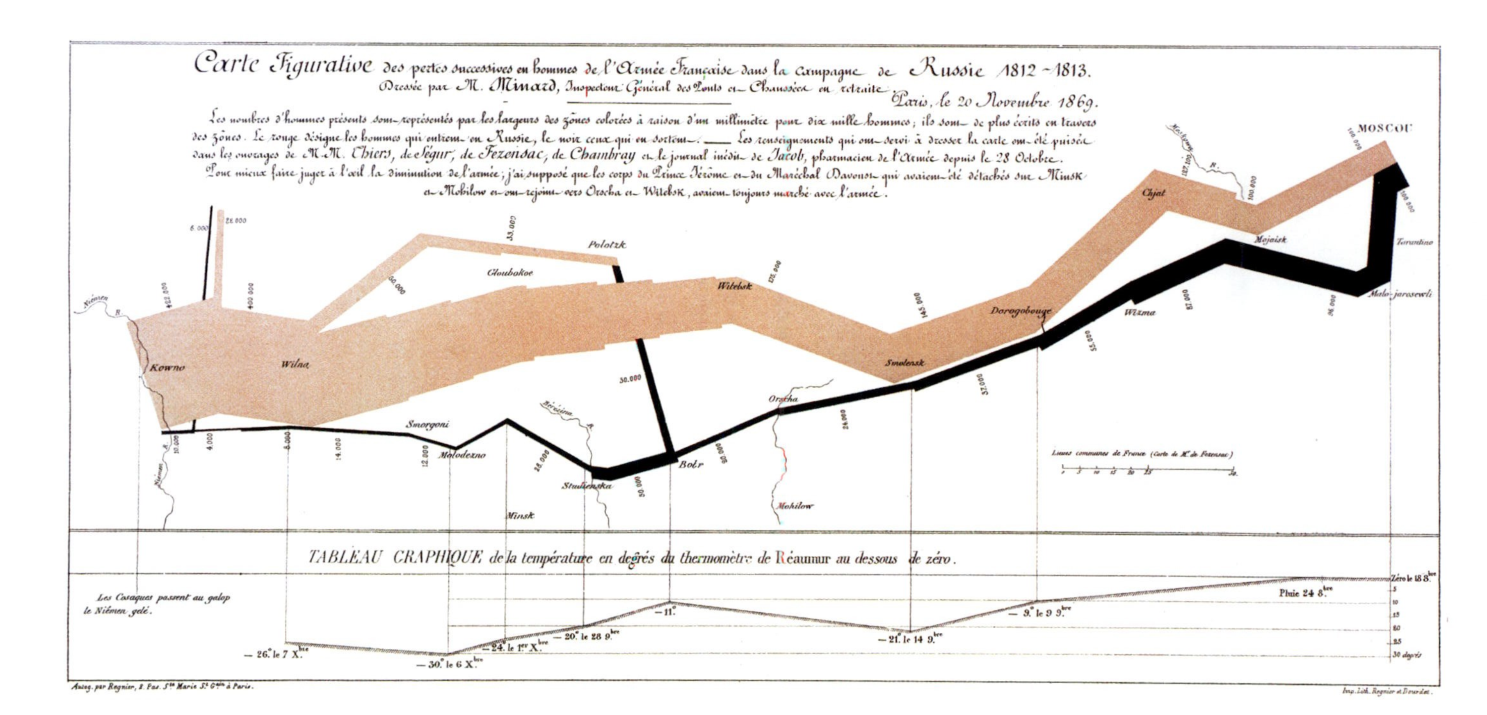


**Figure 18.2.** Charles Joseph Minard's famous plot of Napoleon's ill-fated Russian campaign of 1812 (reprinted courtesy of Graphics Press, Cheshire CT).

Figure 15.3. Francis Galton's "Isochronic Passage Chart for Travellers." The colors indicate the number of days it would take to travel to each destination from London in 1883. Stephen Stigler provided this copy to us. We are grateful to him for his generosity.

Figure 15.3. (*Continued*)

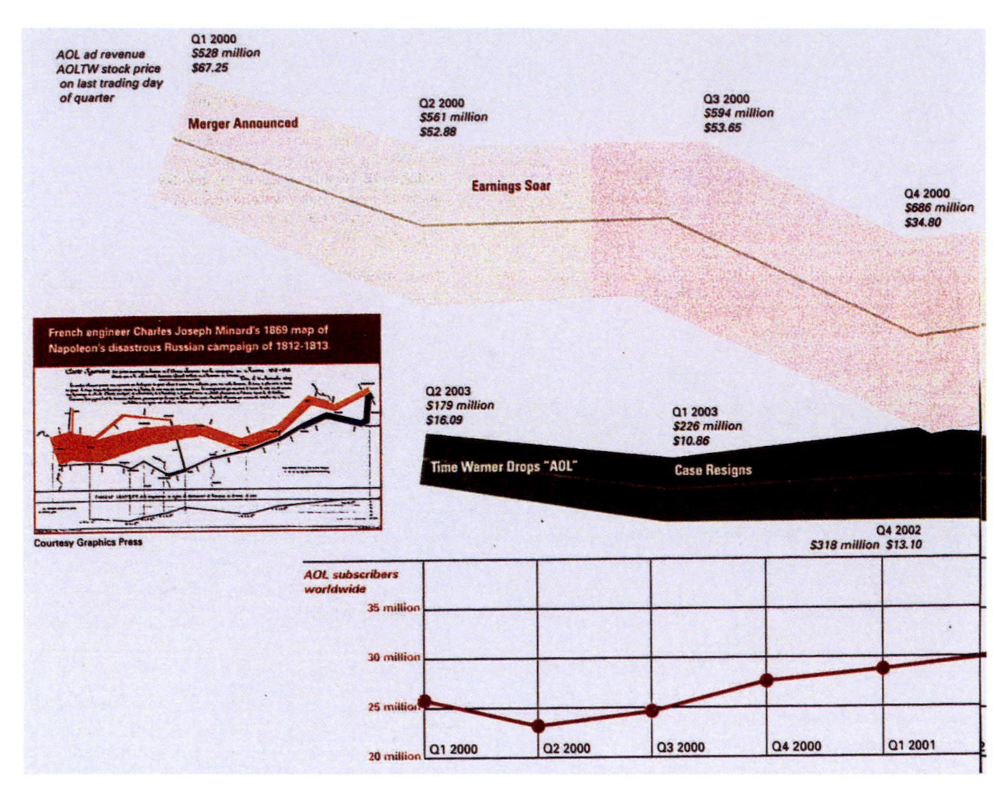


Figure 18.3. AOL's death march. Produced by Tom Stein in *Wired*, November 2003, pp. 62–63.

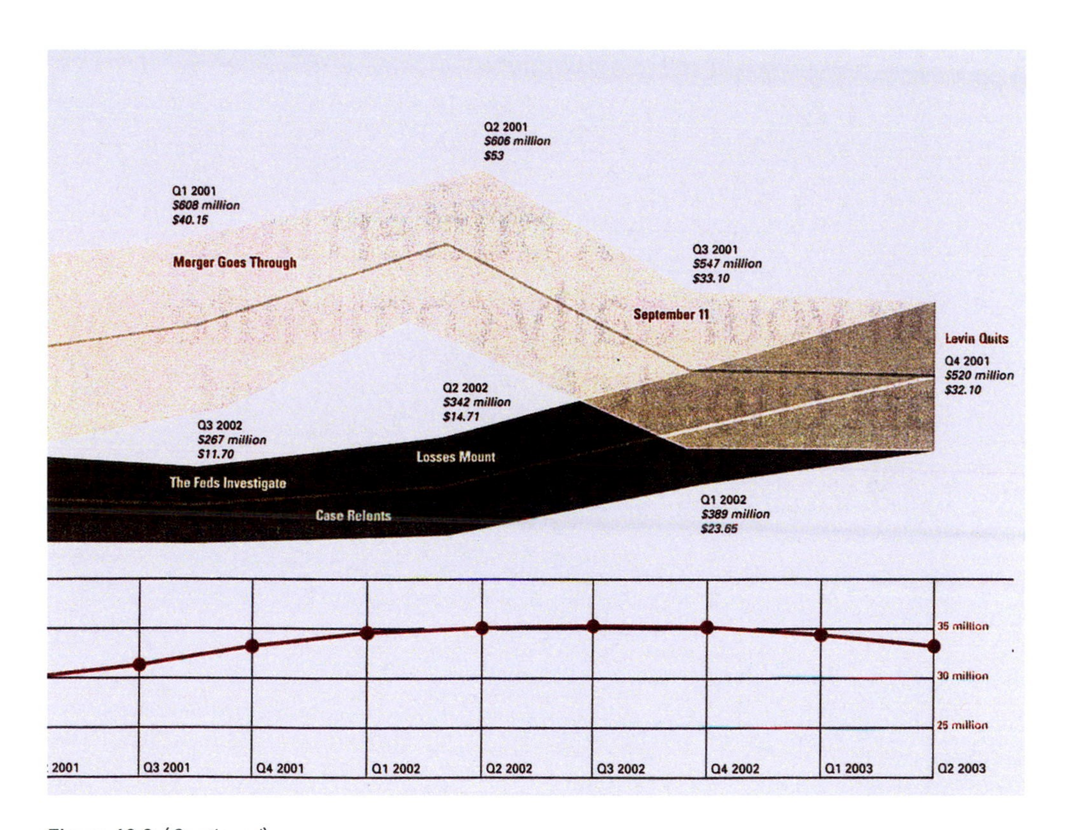


Figure 18.3. (*Continued*)

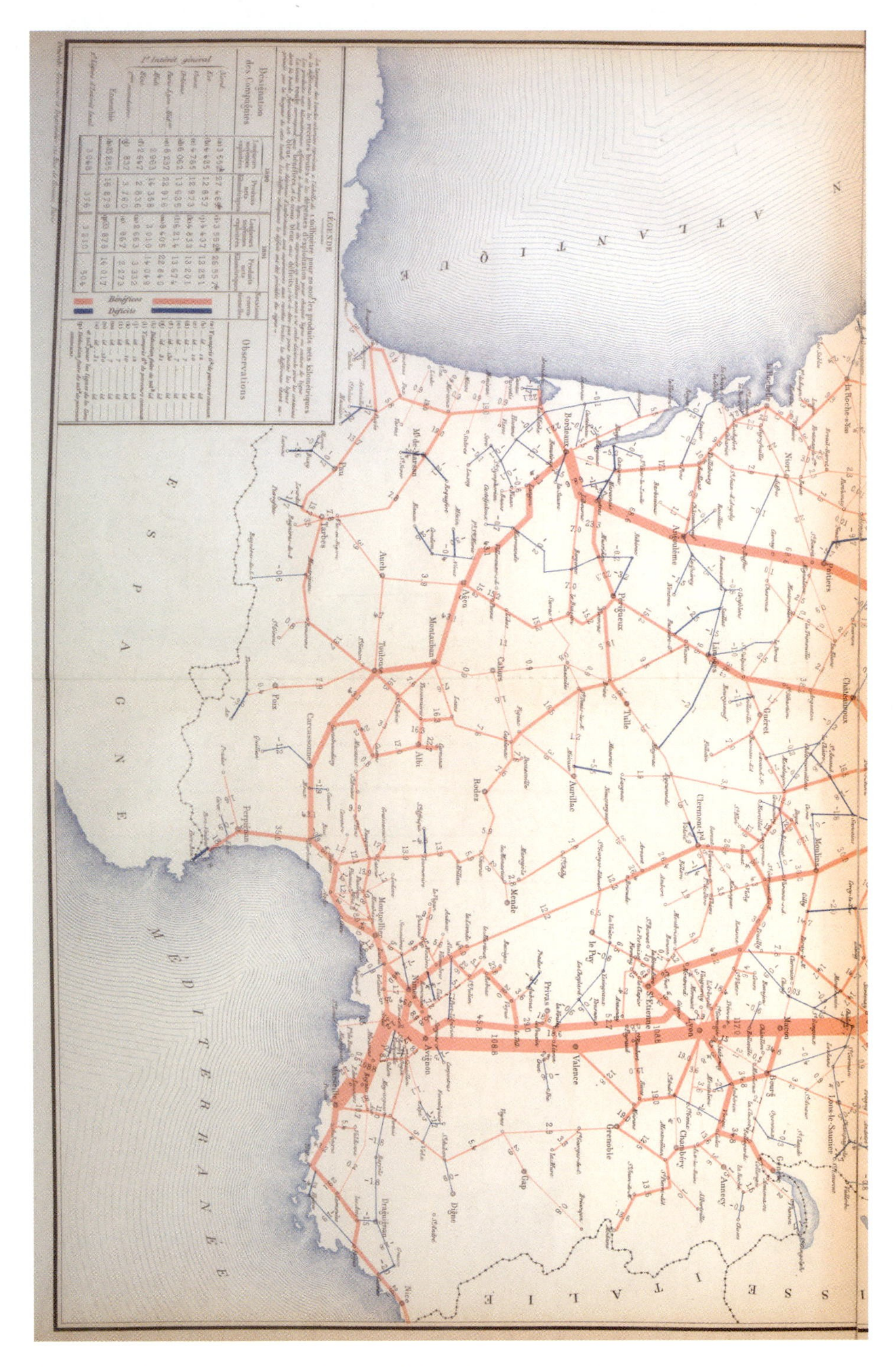


Figure 19.1. The net profits per kilometer of French railways in 1891.

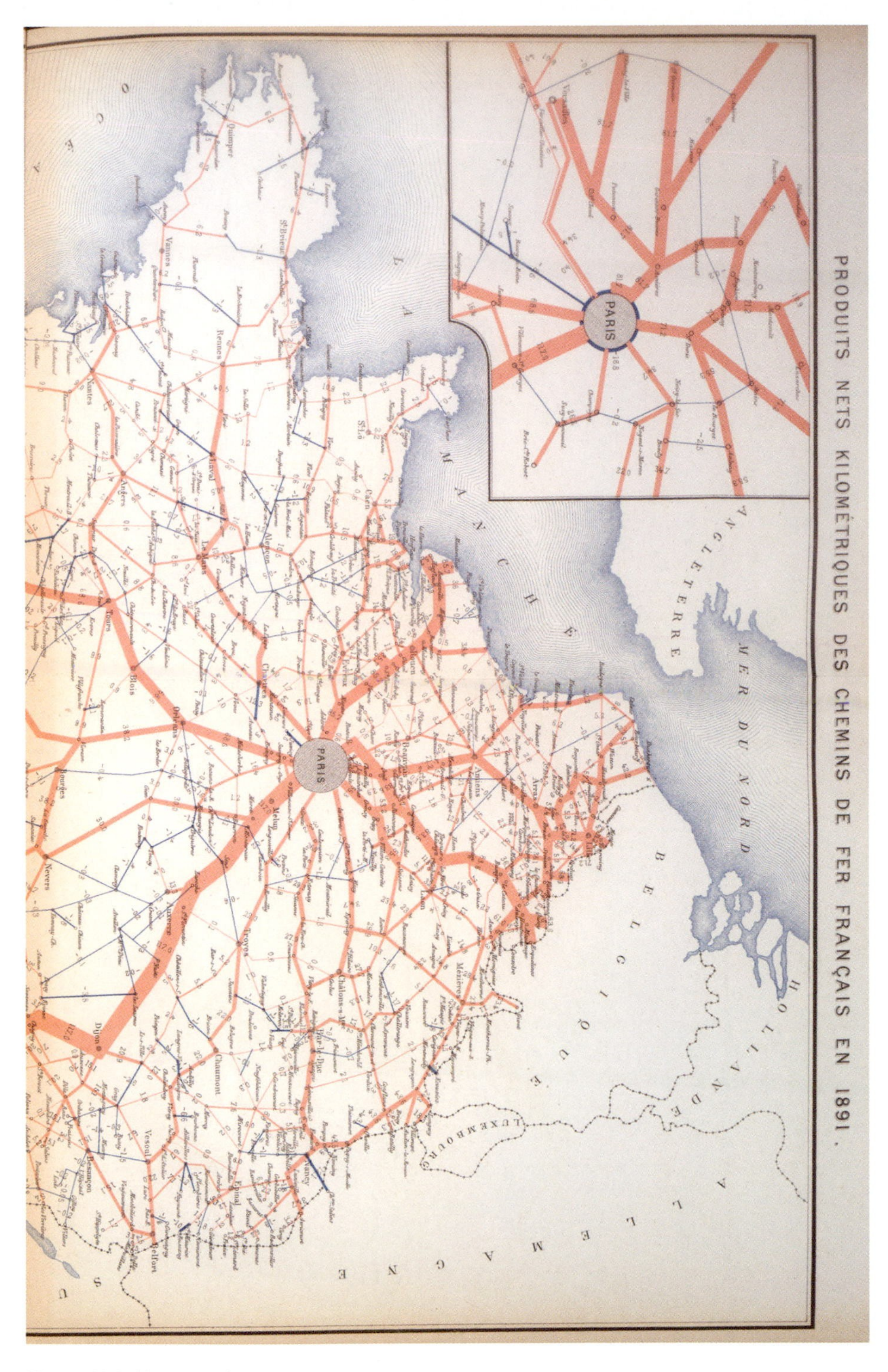


Figure 19.1. (*Continued*)

Figure 19.2.
Average price for transportation of English coal in 1892.

/ C12

Figure 19.2.
(*Continued*)

6

אין סעפטעמבער 148 אף 100 מענער און אין
נאוועמבער 132 " 100 " .

ווי ווייט די פארהעלטענישן ווייכן אפ פון דער נארמע באווייזן די פארגלייכן מיט
אייראפיישע לענדער. די העכסטע פראפארץ צווישן מענער און פרויען האט באטראפן 114 פרויען אף
100 מענער (אין עסטלאנד און אין לעטלאנד). די קלענסטע צאל אין אייראפע (אין האלאנד)
איז געווען 101 פרוי אף 100 מענער (1934).

די בייליגנדע דיאגראמע גיט א גראפיש בילד פון ביידע ציילונגען.

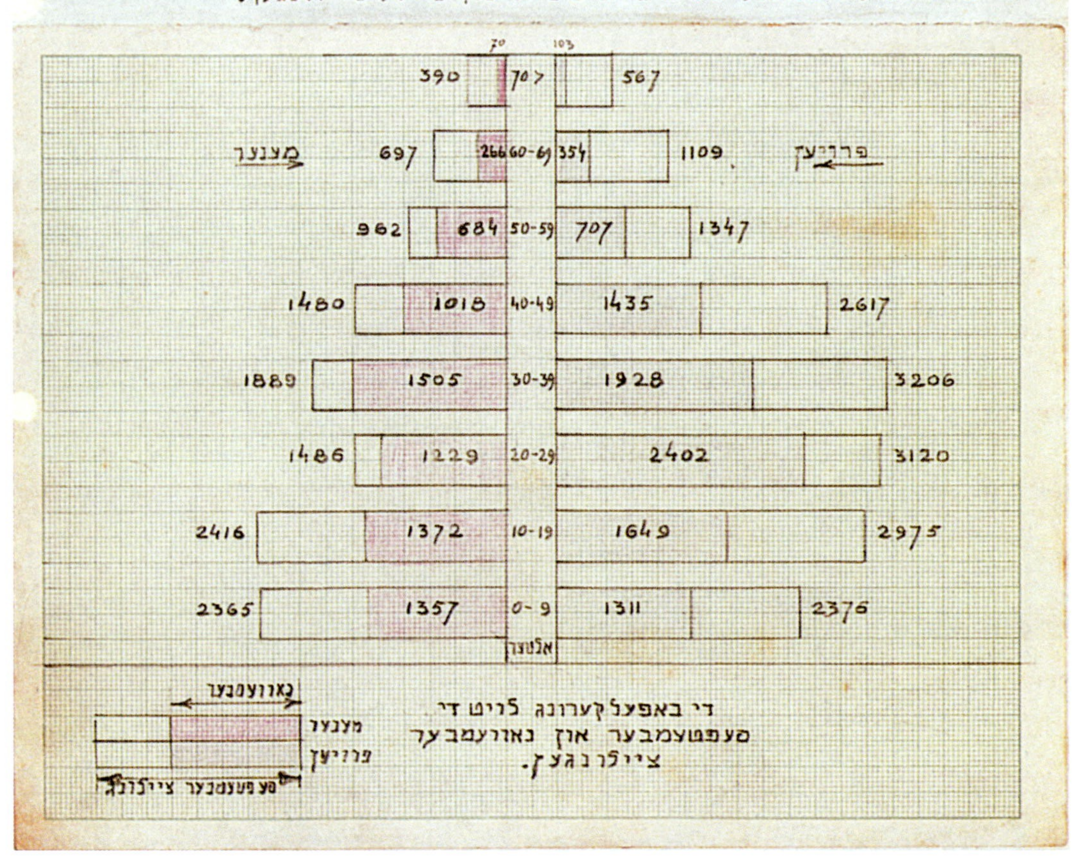


**Figure 21.2.** The population losses in the Kovno Ghetto due mainly to the "Great Action" of October 28, 1941. Males are represented on the left, females on the right. The shaded portion represents those still surviving in November. The central column shows the age groups depicted. (Page 69.)

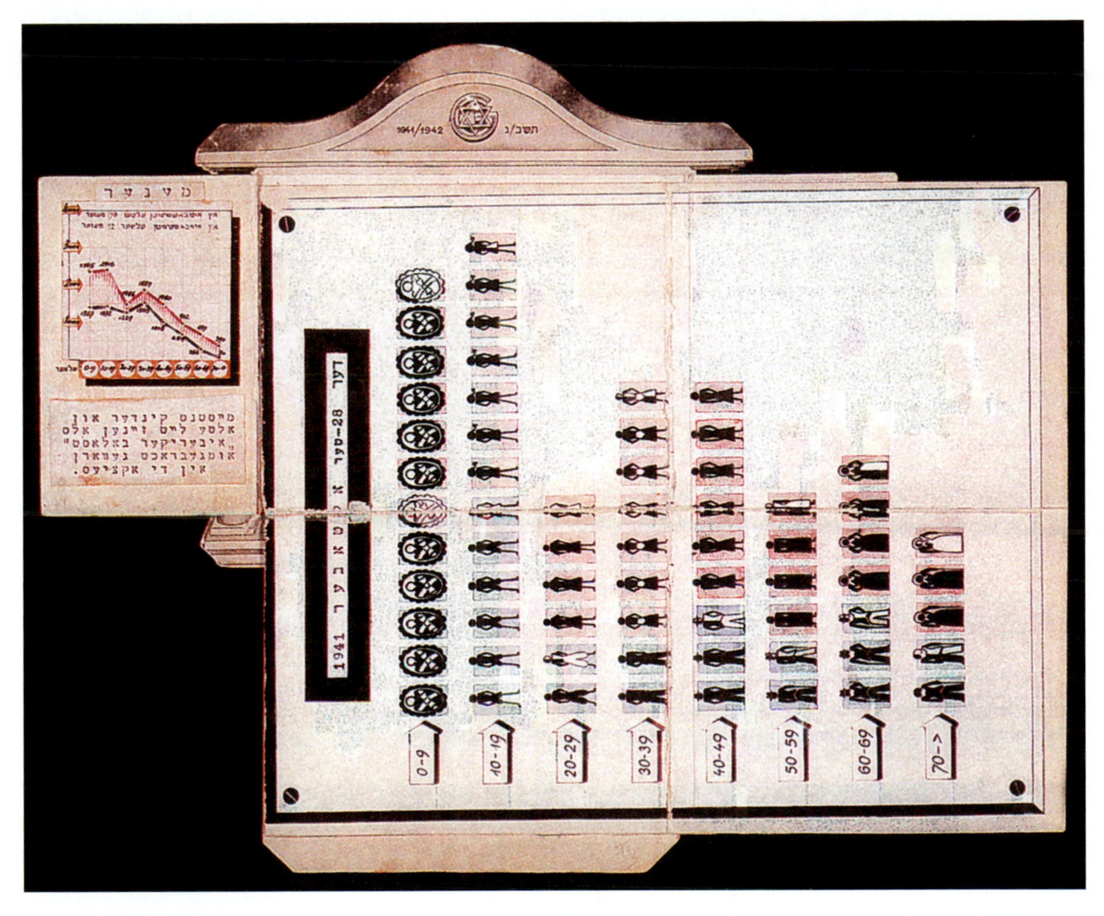


**Figure 21.3.** The same data as shown in figure 21.2 in a different form. Each icon represents about 200 of the victims. (Page 160.)

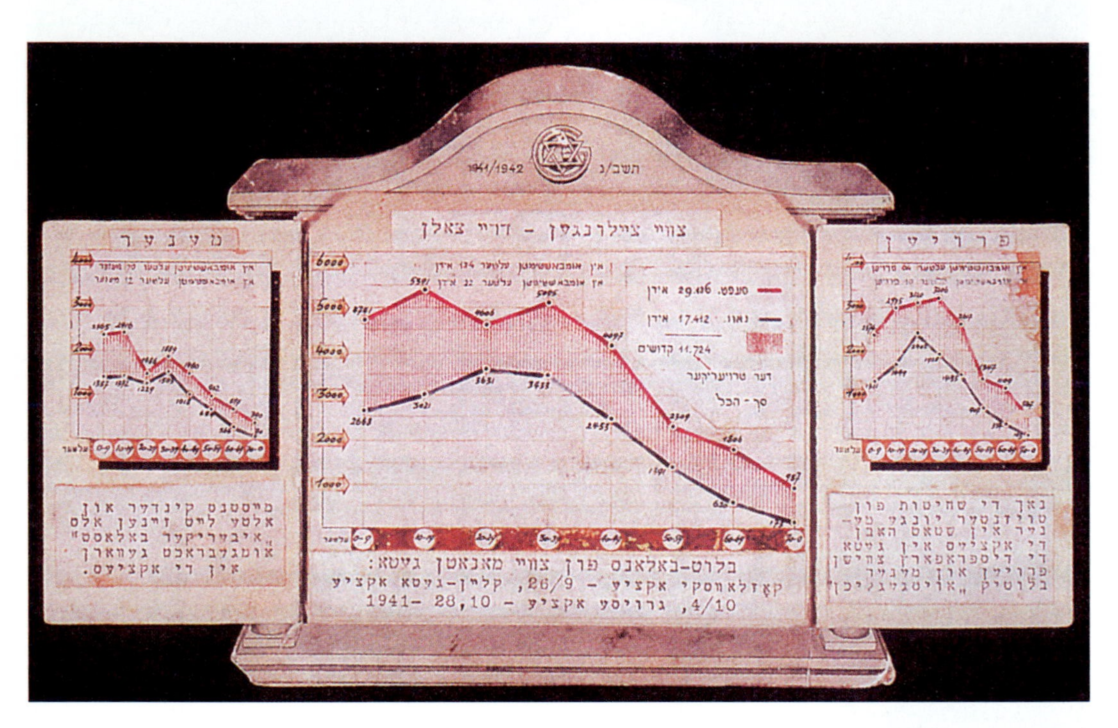


**Figure 21.4.** The same data as in figures 21.2 and 21.3, shown this time as line charts, both total (in the center) and separately for men (on the left) and women (on the right). The upper lines represent the age distribution of the ghetto population in September, the lower line in November. The shaded portion then represents those murdered between those two dates. (Page 158.)

theory almost perfectly. In fact, in 1936, when a statistical test of goodness of fit to the genetic hypothesis was performed, it indicated that the data were fit better than a reasonable person should expect. This led the famous British statistician Ronald Fisher[1] to conclude that, although Mendel had correctly described the experimental layout, the results were, literally, too good to be true. More than thirty years later, van der Waerden[2] inferred that the too-good-fit was not due to any dishonesty on Mendel's part, but rather to the arbitrary stopping rule that he was thought to have employed; he ran the experiment and kept track of the proportion of pea types until it matched his expected value, and then he stopped.

It is well known that when you make decisions about the design of an experiment on the fly, on the basis of the data being generated, great care must be exercised. But sometimes such decisions are subtle—we may make them without noticing. This chapter is a description of one such insidious situation.

## 14.2. THE PROBLEM ARISES

In any test administration things can go wrong. In traditional paper and pencil testing the pencil point can break, the paper can tear, test forms can be misprinted, answer sheets can be mismatched with the test form. As more technology is used some problems disappear and others rise to take their place. Running out of scrap paper for calculations is replaced by having calculator batteries fail, and in computer-administered tests a new world of glitches is manifest. As has always been the case, it is incumbent on the test administrator to look into all sources of administration variation and determine the extent to which their deviation from standardization affects the examinees' performance.

As part of such an analysis, we examined the scores obtained on a computer-administered exam as a function of the amount of delay caused by a computer malfunction. The result of this is shown in table 14.1 for 359 examinees who were so affected. The result was both clear and unexpected—the longer the delay, the higher the score.

When these figures are graphed the trend becomes clearer still (figure 14.1), and the evidence of our eyes is reinforced formally by a statistical test on the trend.

TABLE 14.1
**Scores on a Computer-Administered Exam as a Function of Delay Caused by a Computer Malfunction**

| *Range of delay (minutes)* | *Mean score* | *Number of observations (n)* |
|---|---|---|
| 0 to 11 | 212 | 25 |
| 11 to 20 | 214 | 54 |
| 20 to 30 | 216 | 76 |
| 30 to 40 | 217 | 58 |
| 40 or more | 219 | 146 |
| **OVERALL** | **217** | **359** |

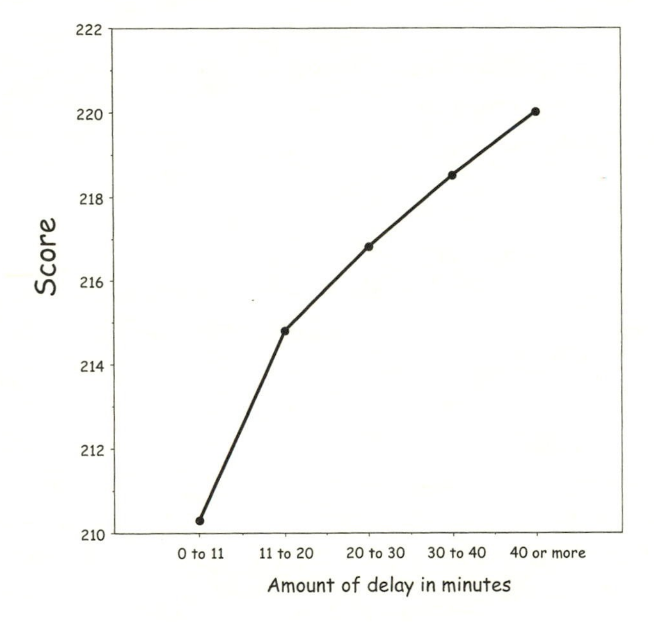


**Figure 14.1.**
As the delay increases so too does the score. Trend is significant beyond .05 level.

Thus arose the mystery. Why should a longer delay yield higher scores? No good explanation suggested itself and so we began to suspect some sort of statistical anomaly. Could the data shown in figure 14.1 be the type I error that we are always worrying about? If it is a statistical anomaly what sort? More than half a century ago Herman Chernoff and Erich Lehmann proved a theorem that had a hint. They showed

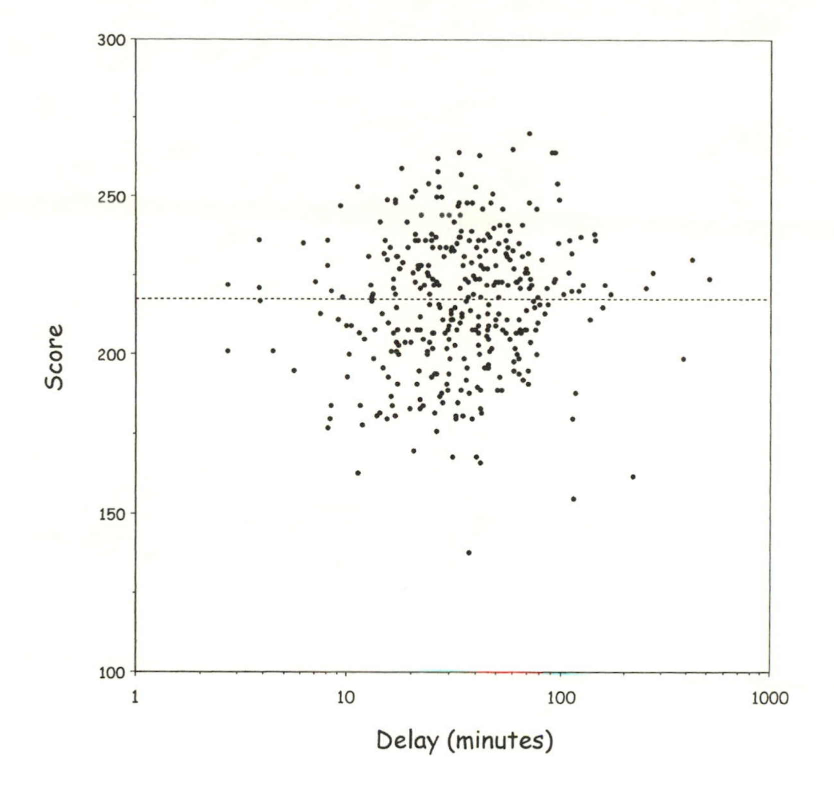


Figure 14.2.
There is no apparent relationship between the length of the delay and the examinee's overall score.

that if you group data into categories and then compared the binned frequencies to some theoretical distribution the resulting comparison statistic is not distributed as they expected. Instead, there is a small bias due to the choice of the bin boundaries. In the situation they examined the bias was small and so the standard approach was a good approximation. But what if we had a particularly unfortunate choice of bin boundaries? How large could the bias become? Could it be large enough to yield substantially incorrect inferences? Perhaps now the connection with baseball announcers and Mendel's peas is becoming more obvious.

To begin to answer these questions I unbinned the results and looked at a scatter plot of the data (figure 14.2). We immediately see a scatter that shows no obvious pattern.

Had we looked at the means in table 14.1 in conjunction with their standard deviations (figure 14.3) we would have realized that

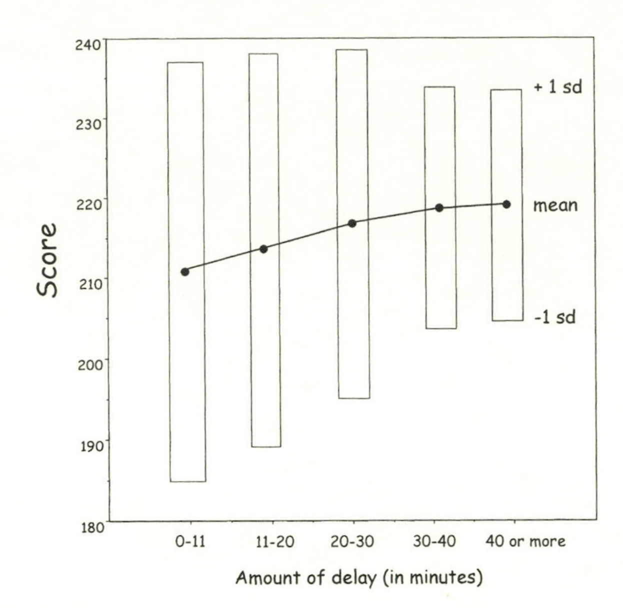


**Figure 14.3.**
Apparent increase is obviously not significant.

the apparent increase in test scores was almost surely an artifact. But how did this come about? What happened?

### 14.3. THE SOLUTION

> With four parameters I can fit an elephant;
> with five I can make it wiggle its trunk.
>
> —John von Neumann[3]

I hinted at the solution in the opening paragraphs of this chapter. If you have the freedom to choose both the number of categories to be used and what their boundaries are to be you can influence the direction of the results. But how much can you influence them? That is the surprise, for with enough freedom you can change a data set with no directional pattern into an increasing function, a decreasing function, or any shape in between. To illustrate, consider the results shown in table 14.2, which reconfigures the category boundaries from the same data shown in table 14.1 but now shows the mean scores to decrease with increasing delay.

Table 14.2
**Reconfiguration of Category Boundaries from Data of Table 14.1 Showing Opposite Conclusion**

| *Range of delay (minutes)* | *Mean score* | *Number of observations (n)* |
|---|---|---|
| 0 to 3.8 | 220 | 4 |
| 3.8 to 8. | 219 | 8 |
| 8 to 113 | 217 | 330 |
| 113 to 170 | 215 | 11 |
| 170 or more | 210 | 6 |
| **OVERALL** | **217** | **359** |

Is there something funny about this particular data set that allows us to reverse the direction of the effect? Or can we always do it? The short answer is that when we have two uncorrelated variables we can always make the data do whatever we want so long as the sample size is large enough.

I will not go through a formal proof, although I will sketch the logic of one. A more detailed proof leans on the law of the iterated logarithm[4] drawn from the literature of Brownian motion and game theory and is too technical for my purposes here.

The basic idea can be easily understood through an example. Suppose we have two unrelated variables, say $x$ and $y$. If we divide the data in half on $x$, we would expect the means in $y$ of the two halves to not be exactly equal. One side will be higher than the other. But if we then move the dividing point first one way and then, if we are not successful, the other way, we will find that this order can usually be reversed. For example, on the data set we have looked at so far, we can divide them into two categories in different ways yielding an upward or downward trend, as we wish.

| *Range of delay (minutes)* | *Mean score* | *Number of observations (n)* |
|---|---|---|
| 0 to 12 | 209 | 29 |
| 12 or more | 217 | 330 |
| | | |
| 0 to 170 | 217 | 353 |
| 170 or more | 210 | 6 |

We can then further divide the sample into three categories getting the same effect.

| *Range of delay (minutes)* | *Mean score* | *Number of observations (n)* |
|---|---|---|
| 0 to 12 | 209 | 29 |
| 12 to 70 | 218 | 281 |
| 71 or more | 222 | 49 |
| | | |
| 0 to 4 | 220 | 4 |
| 4 to 170 | 217 | 349 |
| 171 or more | 210 | 6 |

And divide again into four categories, still being able to yield both increasing and decreasing trends.

| *Range of delay (minutes)* | *Mean score* | *Number of observations (n)* |
|---|---|---|
| 0 to 12 | 209 | 29 |
| 12 to 15 | 210 | 13 |
| 15 to 71 | 216 | 268 |
| 71 or more | 222 | 49 |
| | | |
| 0 to 4 | 220 | 4 |
| 4 to 112 | 217 | 337 |
| 112 to 171 | 212 | 12 |
| 171 or more | 210 | 6 |

An algorithm for finding such results is clear. Suppose we wish to find boundaries such that a five-category system will be monotonically increasing.

We begin at the high end of $x$ and move in until the mean in the thus formed category (1) is two higher than the mean. Then we go to the other end of the distribution and move to the right until the mean within that category (2) is two less than the mean. Next we return to the newly formed boundary on the high end and start the random walk again, moving left until the mean within this category (3) is one greater than the mean. We then repeat the random walk process on the low end until the mean of category (4) is one less than the mean. We have used up all of our degrees of freedom, but it is likely that the mean of the remaining middle category (5) will be close enough to the mean for the resulting result to be monotonic.

| 2 | 4 | 5 | 3 | 1 |
|---|---|---|---|---|

To make it monotonic in the other direction, just reverse the process. How do we know that we can get a mean to be big enough? A random walk in one dimension will eventually reach any value you wish if you let it go long enough. But how long is "long enough"? We know that it follows the law of the iterated logarithm, which means that, as the goal gets further away from the mean, the number of trials it takes goes as a double exponential. Knowing this, it is perhaps surprising that for the modest goal we set for ourselves here, making a monotonic trend in either direction, a sample of 359 was enough. For more categories and for larger trends a much larger sample may be required.

### 14.4. AN ALGORITHM

For the purposes of this chapter my colleague Marc Gesseroli prepared a computer program that will search for category boundaries that yield the best fit to the sort of structure that the user specifies. We call this program RESCUE and is so named for it can be used by researchers who know, deep in their hearts, that they have a real effect even through their data show no evidence of it. RESCUE can turn such null results into significant trends simply through the manipulation of the number and size of the binning categories. It is available free of charge, although he reserves the right to publish the names of all researchers who request it.

### 14.5. THE LESSONS LEARNED

We have seen here that, even with relatively modest sample sizes, there are still enough degrees of freedom in a data sample to allow the binning of the data to yield apparent trends in the binned means—even when there are no such trends in the underlying data. Since we often look at means from binned data, how are we to know if the observed trends are real or epiphenomenal? Obviously, looking at a scatter plot of the unbinned data is the first choice. But what are we to do if the data come prebinned (as in surveys where data are reported by prespecified categories) and you cannot get the original data? This is a difficult problem. One approach to it is simulation. That is, you can try to recreate the data through simulation by generating a plausible data set that matches the binned results' means and variances. Then take this simulated data set and input it into RESCUE and see if the algorithm can produce trends in other directions. If so, and if the chosen bins are not too implausible, you have been warned.

We understand best those things we see grow from
their very beginnings.

—*Aristotle*, Metaphysics

# PART V History

The science of uncertainty has been under development for more than three hundred years.* In this section I pay homage to those who preceded us by using modern tools to investigate ancient puzzles. In chapter 15 we look into two aspects of the literary influence of the British polymath Francis Galton (1822–1911), who among his other accomplishments published several works on the use of fingerprints to identify criminals as well as a marvelous map showing distances from London in days of travel. Galton was an early fan of the tools of the science of uncertainty, and especially of the normal distribution. In chapter 16 we study one of his uses of this distribution and uncover a problem that he should have seen, but missed. Galton's error was in not recognizing how unusual are truly rare events. This also helps to explain why the National Basketball Association scours the world for athletes who might play center, as well as why they are so highly compensated.

As we have seen, the graphic display of quantitative phenomena is a principal tool in understanding, communicating, and eventually controlling uncertainty. Thus it is altogether fitting and proper that we spend some time discussing the developers of these tools. Principal among these is the inventor and popularizer of most of the modern

*Jacob Bernoulli (1654–1705) is usually designated as the father of the quantification of uncertainty (Stigler, 1986, p. 63) whose work was amplified and deepened by Abraham De Moivre (1667–1754), a French expatriate who lived in England (he was elected to the Royal Society in 1697, when he was but thirty).

graphical forms, the remarkable Scot William Playfair (1759–1823). But he did not invent the scatter plot, the most versatile and powerful weapon in the arsenal of all who practice the science of uncertainty. Why? In chapter 17 I explore this question and show how even Playfair's ingenious designs could be improved upon had he but imagined the scatter plot.

The more than two centuries that have passed since Playfair's graphs appeared have yielded very few real improvements. The discussion of the scatter plot is relevant exactly because Playfair didn't invent it.* Charles Joseph Minard, once dubbed, the "Playfair of France," produced a body of graphical work that is breathtaking. One of his graphs, showing Napoleon's ill-fated 1812 incursion into Russia, has been called the greatest graph ever drawn. It is natural to expect others to borrow from such a master. In chapter 18 I discuss one such counterfeit and, by placing it alongside the original, illustrate its unworthiness. I also show alternative ways that the same data could have been shown to better advantage. Then in chapter 19 I describe some other graphs produced by other nineteenth-century followers of Minard that are worthy and that helped their users navigate the uncertainty of their times.

Twentieth-century France has yielded yet another graphical master, Jacques Bertin, whose *Semiologie Graphique* is remarkable because it serves both as a practical handbook for those who wish to design visual aids to navigate uncertainty, and as a deep theoretical discussion of the structure of *La Graphique* and how good design interacts with human perception. In chapter 20 I recount the contents of a letter received from Bertin, written in his eighty fifth year, describing some of his latest insights. My hope is that readers will hunger for a complete meal after tasting this graphical appetizer.

It is part of the human condition to want to leave something of ourselves behind after we depart from this Earth. What we leave varies as widely as we vary. People who can write might leave a literary tombstone, a book, a story, a poem; those that are musically inclined might leave a song; others something more physical; and the truest part of ourselves we leave might be our children. In the final chapter of this

* The significance of Playfair's role in the development of the scatter plot is akin to the role of the hound on the night of the murder in Sherlock Holmes' "Silver Blaze."

book I have chosen to leave you with the story of the Kovno Ghetto. Its inhabitants knew that their murders were imminent and many chose to leave behind something that would provide a picture of what happened. Some, no doubt, wrote songs, some poems, some stories. But some, of a more empirical bent, drew graphs. Chapter 21 contains these graphical depictions of the horror that was the Holocaust.

# 15

## Truth Is Slower than Fiction

The race is not always to the swift,
nor the battle to the strong,
but that's the way to lay your bets.

—Damon Runyan

Modern science often involves detective work in which clues are followed and inferences drawn on the basis of those clues. We are almost never certain that our inferences are correct, but we usually have to take some action, and so we make our best guess. As we have seen in earlier chapters, the circumstantial evidence gleaned from data are often, in Thoreau's memorable words, "like when you find a trout in the milk."*

In this chapter we play historical detective by trying to determine how leading nineteenth-century authors came up with their plot ideas. In our search we use rules of logic well established in classical antiquity, abetted by some modern scholarship. We begin with the most famous

*Thoreau was referring to claims made during a dairyman's strike that some purveyors were watering down the product. (Henry David Thoreau, November 11, 1850.)

detective of them all as he tries to navigate the uncertainty associated with finding a murderer.

> [Lestrade] led us through the passage and out into a dark hall beyond.
>
> "This is where young McFarlane must have come out to get his hat after the crime was done," said he. "Now look at this." With dramatic suddenness he struck a match, and by its light exposed a stain of blood upon the whitewashed wall. As he held the match nearer, I saw that it was more than a stain. It was the well-marked print of a thumb.
>
> "Look at that with your magnifying glass, Mr. Holmes."
>
> "Yes, I am doing so."
>
> "You are aware that no two thumb-marks are alike?"
>
> "I have heard something of the kind."
>
> "Well, then, will you please compare that print with this wax impression of young McFarlane's right thumb, taken by my orders this morning?"
>
> As he held the waxen print close to the bloodstain, it did not take a magnifying glass to see that the two were undoubtedly from the same thumb. It was evident to me that our unfortunate client was lost.
>
> "That is final," said Lestrade.
>
> —Arthur Conan Doyle, 1903,
> "The Adventure of the Norwood Builder,"
> *Collier's Weekly*

"The Adventure of the Norwood Builder" first appeared in *Collier's Weekly* in October of 1903, and in it the, apparently, new technology of using fingerprints to identify wrongdoers was featured. Where had Holmes heard about fingerprints? One plausible place was from Francis Galton's monographs on the topic a decade earlier.[1] Holmes repeatedly refers to his own monograph on the identification of cigarette ashes, but is curiously silent on Galton's work.

Francis Galton was a British polymath who made fundamental contributions to a broad range of areas including weather maps,[2] fingerprints, psychological measurement, and regression to the mean. His work on fingerprints came as a follow-on to the work of the French criminologist Alphonse Bertillon (1853–1914).*

The sequence of "Galton first, Holmes later" supports the natural assumption that writers of fiction keep an eye on the scientific literature, so that they may incorporate the latest developments into their stories. It may be natural, but in this case at least, it would be wrong.

> When I was a youth, I knew an old Frenchman who had been a prison-keeper for thirty years, and he told me that there was one thing about a person which never changed, from the cradle to the grave—the lines in the ball of the thumb; and he said that these lines were never exactly alike in the thumbs of any two human beings. In these days, we photograph the new criminal, and hang his picture in the Rogues' Gallery for future reference; but that Frenchman, in his day, used to take a print of the ball of a new prisoner's thumb and put that away for future reference. He always said that pictures were no good—future disguises could make them useless; 'The thumb's the only sure thing,' said he; 'you can't disguise that.' And he used to prove his theory, too, on my friends and acquaintances; it always succeeded. (Mark Twain, 1883, *Life on the Mississippi* [chapter 31])

There were other, nonfictional, uses of fingerprints before Galton's. In 1882 the U.S. geologist Gilbert Thompson used his own fingerprints on a document to prevent forgery. Still earlier, a British physician, Dr. Henry Faulds, recognized the unique nature of fingerprints and designed his own system of classification. Interestingly, in 1880 he sent a sample of these to Charles Darwin for comment. Darwin, then aged and in ill health, promised to pass the materials on to his cousin, Francis Galton. There were earlier uses of fingerprints. For example, Sir William Hershel, Chief Magistrate of the Hooghly district

*Bertillon was chief of the identification bureau in Paris, where in 1882, he devised a system of identifying individuals though the use of a combination of anthropometric measurements, including height, weight, the left foot, the right forearm, and head measurements. These were combined with photographs to form a record of identification later known as "Bertillonage." Bertillon's photographs were the beginning of modern day "mug shots" (Stigler, 1999, chapter 6). He was the son of the well-known statistician Louis Adolphe Bertillon (1821–1883). Much of his work was made moot by Galton's development of accurate ways of comparing fingerprints.

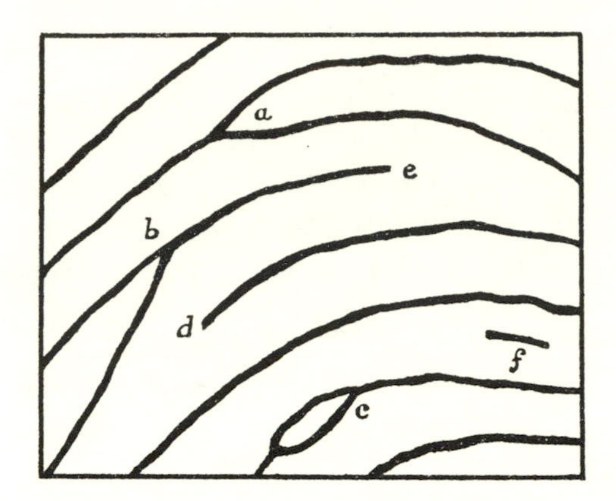


Characteristic Peculiarities in Ridges
(about 8 times the natural size)

**Figure 15.1.**
Galton's illustration of the characteristics of fingerprints, showing six principal types of minutiae (from Pearson, 1930, p. 178, plate 31).

in Jungipoor, India, used fingerprints on native contracts as early as 1858. As his collection grew so too did his suspicion that fingerprints were both unique to the individual and constant over the course of one's life. In 1823, John Purkinji, a professor of anatomy at the University of Breslau, published a discussion of nine different fingerprint patterns, but he made no mention of the value of fingerprints for personal identification. But no one before Galton rigorously looked into their uniqueness, their constancy over time, or any systematic way of classifying them.

Galton, using his own data, as well as Hershel's, provided strong evidence that the characteristics of fingerprints that he would use in his classification system did not change over time, from early childhood to death. His system of classification was based on the minutiae of the prints—the tiny islets and forks. The gross features, the loops and whorls, were only useful for rough classification[3]; it was only through the use of the minutiae that precise identification was possible. Figure 15.1 reproduces a graphic from Galton's work identifying six specific characteristics of finger prints. Figure 15.2 provides an example of several of these same characteristics on a print of Galton's right forefinger.

The conclusion we draw from this example is that, at least for fingerprints, fiction preceded science, at least rigorous science, although it

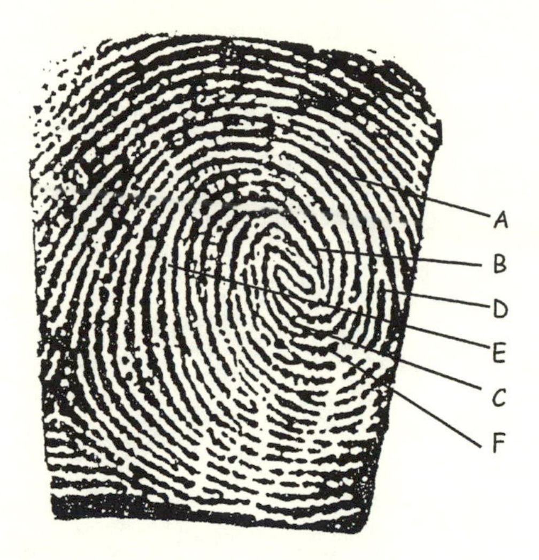


Figure 15.2.
The print of Francis Galton's right forefinger annotated to match Galton's markers in figure 15.1 (from Pearson, 1930, p. 138. figure 14).

was clear that the idea was in the wind before Mark Twain jumped on it. Interestingly, the fictional uses of fingerprints suggest, incorrectly, that just by pulling out his lens and examining the swirls and whirls of the print on the wall Holmes and Watson could tell that it was undoubtedly the unfortunate Mr. McFarlane's.* Apparently the details of classifying a fingerprint were too complex to make for good fiction.

Francis Galton's imagination and industry spanned many other arenas. About a decade before his work on fingerprints he published a marvelous map of the world (figure 15.3) which he called an "Isochronic Passage Chart for Travellers," a map of the world colored according to the length of time it took to travel there from London, in Galton's words, by scheduled "conveyances as are available without unreasonable cost."† (See pages C6–C7 for the color version.)

A quick glance at Galton's map tells us that, in 1883, even the furthermost reaches of the globe could be reached in about forty days. It doesn't stretch one's imagination too much to conclude that it might be within human reach to circumnavigate the planet in eighty days. Could this be where Jules Verne got the idea? Why else would Verne,

* Actually, the print eventually exonerated Mr. McFarlane. Holmes noted that it had mysteriously appeared during the time that Mr. McFarlane had been safely ensconced in jail.

† This map was brought to our attention by Stephen Stigler in his opening address to the fifty fifth Session of the International Statistical Institute, "The Intellectual Globalization of Statistics," and, more or less simultaneously, by Michael Friendly.

Figure 15.3.
Francis Galton's "Isochronic Passage Chart for Travellers." The colors indicate the number of days it would take to travel to each destination from London in 1883. Stephen Stigler provided this copy to us. We are grateful to him for his generosity. (For the color version, go to pages C6–C7.)

a Frenchman, choose Phileas Fogg, an Englishman, as his protagonist? And why would he decide to start and end his journey in London if not for the convenient calculations provided for him by Francis Galton?

Unfortunately, the facts get in the way of this very fine story. Jules Verne published his *Tour du Monde en Quatre-Vingts Jours* in 1872, with George Makepeace Towle's English translation appearing in 1873, fully ten years before Galton's map. While it is always difficult, and often impossible, to know what instigated someone to action, it is not far-fetched to believe that Galton read Verne's story and wondered if it was really possible. Galton was a man of curiosity, imagination, and energy. He was also an ardent empiricist and a firm believer in the power of graphic presentation. His curiosity about the tale might have led him to make inquiries out of which grew this wonderful map.

**Figure 15.4.**
Fingerprints of Sir William J. Hershel's right forefinger at 54 years' interval, the longest known proof of persistence. The 1913 print shows the creases which develop with old age (from Pearson, 1930, opposite p. 438).

We have before us two instances of scientific work by a man of prodigious genius, but in both cases this work was preceded by what appears to be a remarkably prescient work of fiction. Does this mean that Mark Twain and Jules Verne could see further into the future than Francis Galton? While this is certainly a possible conclusion, we think not. Rather it merely shows that doing science is harder and can't be rushed. Figure 15.4 shows the stability of the pattern of Hershel's forefinger print across the fifty four years from 1859 to 1913. It is easy to write something about using fingerprints to identify criminals, but it is a great deal more difficult to explain exactly how to do it and collecting the evidence can take a lifetime.

This tale illustrates an important component of science; it is hard to know new things. The world can yield its secrets to our inquiries, but it almost always takes effort and time.

# 16

## Galton's Normal

The adventure began innocently enough with an email from the Columbia University statistician Andrew Gelman, in which he enclosed a graph that Francis Galton published in his 1869 book *Hereditary Genius*. The graph, shown here as figure 16.1, purports to contain the heights of one million men.

Galton's explanation of the graph was,

> Suppose a million of the men stand, in turns, with their backs against a wall of sufficient height, and their heights to be dotted off upon it. The board would then present the appearance shown in the diagram. The line of average height is that which divides the dots into two equal parts, and stands, in the case we have assumed, at the height of sixty-six inches. The dots will be found to be ranged symmetrically on either side of the line of average, that the lower half of the diagram will be almost a precise reflection of the upper. Next, let a hundred dots be counted from above downwards, and let a line be drawn below them. According to the conditions, this line will stand at the height of seventy-eight inches. Using the data afforded by these two lines, it is possible, by the help of the law of deviations from an average,*

*The normal distribution; Galton cites Quetelet's tables as invaluable aids in using this law. The normal curve has the familiar bell shape and is frequently used as the standard of comparison with other distributions. Distributions that have higher peaks than the normal

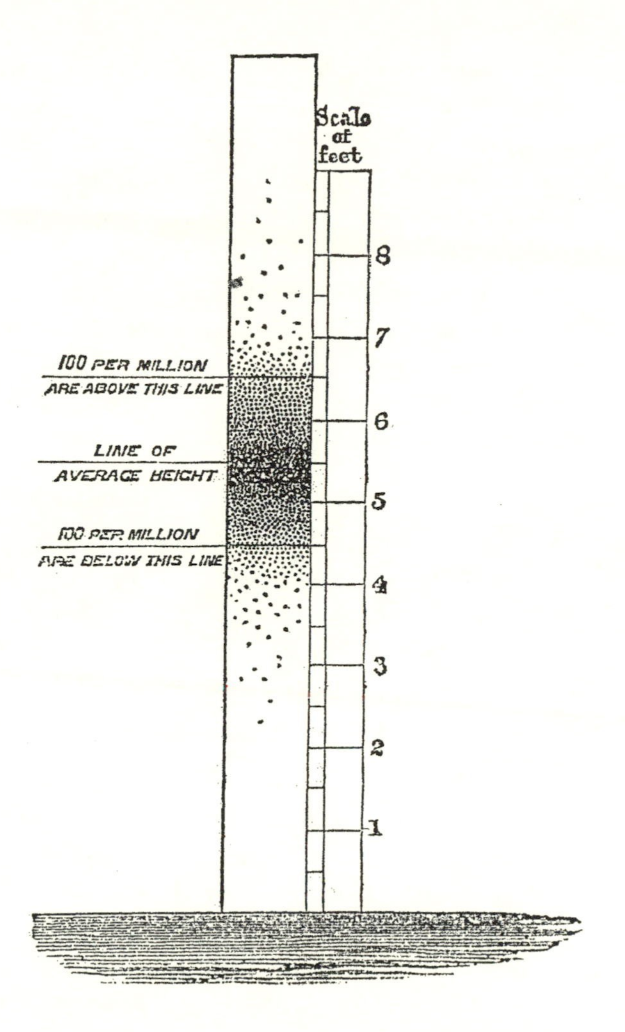


**Figure 16.1.** Galton's hypothetical plot of the heights of one million men (from Galton, 1869).

to reproduce, with extraordinary closeness, the entire system of dots on the board.

Gelman noticed that the tallest person in Galton's graph was about nine feet tall. He said, "it doesn't make sense to me (that) according to the graph, about 1 in a million people is 9 feet tall! There were about

---

are called *leptokurtic;* those with a lower peak and consequently fatter tails are called *platykurtic.* A platykurtic distribution is one that depicts unusual events as occurring more often than would be the case were the distribution normal.

10 million men in England in Galton's time, then that would lead us to expect 10 nine-footers! As far as I know, this didn't happen,* and I assume Galton would've realized this when he was making the graph." Gelman asked if anyone "happened to know the story behind this?"

A step toward easing the mystery was provided by Antony Unwin, whose book (with Martin Theus and Heike Hoffman) *Graphics of Large Datasets: Visualizing a Million* contained the graph that piqued Andrew's curiosity, when he reminded us that these were not real data; Galton was describing a hypothetical population. It was then more completely unraveled by Stephen Stigler, author of a remarkable history of statistics,[1] who observed that,

> he (Galton) specifies for the illustration that the mean is 66 inches† and that 100 out of a million exceed 78 inches. By my rough calculation that gives a standard deviation of about 3.2 inches. This was his earliest statistical book and Galton had more faith in the normal than later, but without good tables available (even though Laplace had given a continued fraction that would have given acceptable results) Galton did not appreciate how fast the tail comes down at the extremes.

Redoing Stigler's calculations a little more slowly, we see that we must calculate the standardized ($z$) score associated with a probability of 100 out of 1,000,000 to be above 78 inches. The $z$-score for 1/10,000 is 3.72 and so we calculate the standard deviation to be $(78 - 66)/3.72 = 3.2$. We then ask how many standard deviations away from the mean is it for a nine-footer. The obvious calculation $[(108 - 66)/3.2]$ yields a $z$-score of 13.13. If height truly was distributed normally, the likelihood of a nine-footer would be far, far less than one in a trillion.‡ This is

*Robert Pershing Wadlow (1918–1940) at 8 feet 11 inches is the tallest human on record. When he died he was still growing and it is not implausible that he would've broken the nine-foot barrier. He was thought to be a "pituitary giant" whose unusual stature was due to an overactive pituitary, and so should not be considered as part of Galton's "normal" population.

†This figure is smaller than that from contemporary surveys (Rosenbaum et al., 1985) for both the mean for British men (69.2 inches) and the standard deviation (2.8 inches).

‡Doing the same calculations suggests that eight-footers are 9.4 standard deviations (SDs) above the mean (still very unlikely to be found), but seven-footers who are only 5.6 SDs above the mean are about ten in a billion and hence we should not be surprised that professional basketball teams, after scouring the world, have found a few. From this we can also calculate the expected height of one person in a million (81.3 inches) as well as the expected

what led to Stigler's inference that Galton didn't fully "appreciate how fast the tail comes down at the extremes."

Until I did these calculations, neither did I.

*An Explanatory Addendum.** The metric of probabilities (the area under the normal curve after a particular point) is sometimes hard to comprehend, especially for very small probabilities. But since the area under the curve in a specific region is directly related to the height of the curve in that region, it is a little easier if we consider the relative height of the normal curve. If the height of the normal curve 13 standard deviations away from the middle was 1 mm, let us calculate what the height of the same curve would be at the middle. To do this we use the equation of the normal curve, which yields

$$\text{When } z = 13, \quad \frac{1}{\sqrt{2\pi}} \exp\left(-\frac{1}{2}z^2\right) = 8 \times 10^{-38} \text{ approximately.}$$

$$\text{When } z = 0 \text{ (at the middle)}, \quad \frac{1}{\sqrt{2\pi}} \exp\left(-\frac{1}{2}z^2\right) = 4 \times 10^{-1}$$

approximately.

To find the scaled height of the curve at the mode, we must solve the proportionality problem

$$8 \times 10^{-38} \text{ is to } 1 \text{ mm as } 4 \times 10^{-1} \text{ is to } x \text{ mm.}$$

After solving this, we find that $x$ is approximately equal to $5 \times 10^{36}$ mm or $5 \times 10^{30}$ km.

Note that a light year is $9.5 \times 10^{12}$ km, which means that the curve at its middle would be $5.3 \times 10^{17}$ light years high. But the universe is estimated to be only 156 billion light years across.

This means that the height of the normal curve at $z = 0$ would have to be 3.4 million times larger than the universe.

---

height of the tallest person among the eleven million males living in England at the time—82.8 inches. Another example of Mark Twains observation about the joy of science,—with such a small investment in fact we can garner such huge dividends of conjecture.

*This expansion follows a suggestion of Eric Sowey, a professor of statistics at the University of New South Wales. I am grateful for his aid in helping me to more fully understand how quickly the normal distribution drops off.

This makes explicit just how difficult it would be to try to draw the standard normal distribution to scale on paper under these circumstances.

Aside: if the height of the normal curve was 1 mm at $z = 6$ the height at $z = 0$ would be 66 km. Thus it still could not be drawn to scale.

# 17

## Nobody's Perfect

One of the great advantages of the graphical depiction of information is the vast flexibility of graphical formats. The same format can be used for many different purposes. Nevertheless, line charts, bar charts, and pie charts seem to be most frequently used for presentation of quantitative information in the popular media, whereas the scatter plot's principal home is on the pages of more technical outlets. But although its fame may be a bit circumscribed, the scatter plot's value within science is fully appreciated. It has been described[1] as the most versatile, polymorphic, and generally useful invention in the entire history of statistical graphics.

In view of the division between typical audiences of these two classes of displays (scatter plots vs. everything else) it should not be surprising that there were two different inventors. The iconoclastic Scot William Playfair (1759–1823) is generally credited with the invention of the pie chart, bar chart, and line chart,[2] whereas the British astronomer John Frederick William Herschel (1792–1871) is the most likely candidate as the inventor of the scatter plot.[3] Herschel's use of the scatter plot occurs in his "investigation of the orbits of revolving double stars," which he read to the Royal Astronomical Society on January 13, 1832, and published a year later. The term "scatter plot" appeared somewhat later; and is generally credited to Karl Pearson.[4]

But why didn't Playfair invent the scatter plot? He did such a marvelous job inventing so many graphical tools, how could he have missed

Plate 23

Exports and Imports of SCOTLAND to and from different parts for one Year from Christmas 1780 to Christmas 1781.

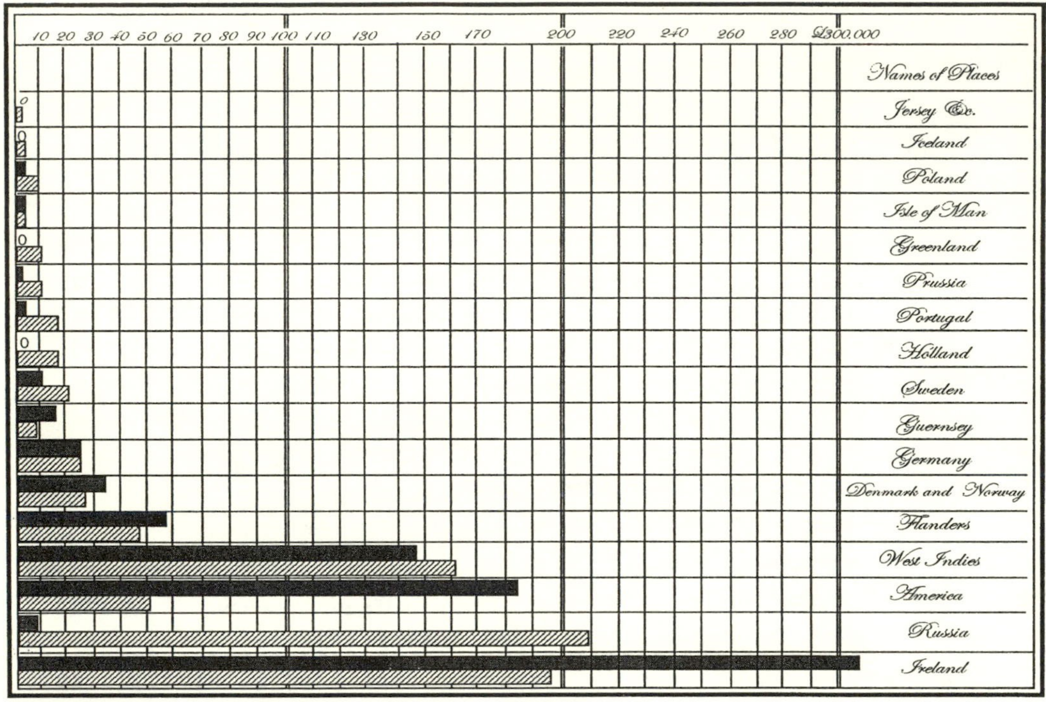


The Upright divisions are Ten Thousand Pounds each. The Black Lines are Exports the Ribbed lines Imports.

Published as the Act directs, June 7th. 1786 by Wm. Playfair

Neele sculp. 352 Strand, London

**Figure 17.1.** The imports (cross-hatched lines) and exports (solid lines) to and from Scotland in 1781 for eighteen countries, ordered by the total volume of trade.

this one? Could it be because the data of greatest interest to him would not have yielded deeper insights as scatter plots? Playfair's initial plots (in his 1786 Atlas) were almost all line graphs. He was interested in showing trends in the commerce between England and its various trading partners. Hence, line graphs of imports and exports over time were the logical design of choice. He would also shade the area between the two resulting curves to indicate the balance of trade, and label this space as balance in favor or against England as the case might be. Such time series data are not a fertile field from which might grow a scatter plot.

Playfair did not have Scotland's trade data for more than a single

Exports and Imports of SCOTLAND to and from different parts for one Year from Christmas 1780 to Christmas 1781.

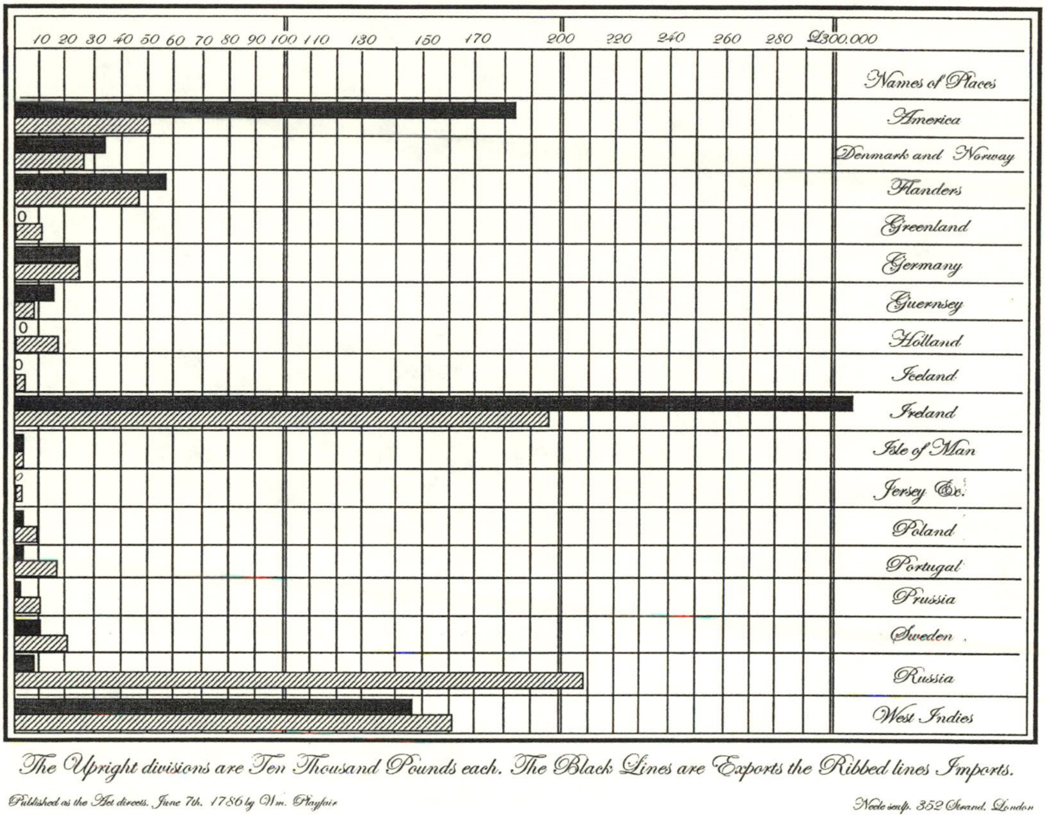

10 20 30 40 50 60 70 80 90 100 110 130 150 170 200 220 240 260 280 £300,000

Names of Places
America
Denmark and Norway
Flanders
Greenland
Germany
Guernsey
Holland
Iceland
Ireland
Isle of Man
Jersey &c.
Poland
Portugal
Prussia
Sweden
Russia
West Indies

The Upright divisions are Ten Thousand Pounds each. The Black Lines are Exports the Ribbed lines Imports.

Published as the Act directs, June 7th. 1786 by Wm. Playfair

Neele sculp. 352 Strand, London

**Figure 17.2.**
The imports (cross-hatched lines) and exports (solid lines) to and from Scotland in 1781 for eighteen countries, ordered alphabetically.

year (1780–1781). Thus the time-line plot was not suitable. So instead, he depicted Scotland's trade with its eighteen partners as a bar chart (figure 17.1). He complained about the insufficiency of the data, but note how well he used the form: he paired imports and exports together for each of Scotland's trading partners for easy comparison (he could have made two separate charts), he placed the bars horizontally to make reading of the labels easier, and he ordered the countries by the total value of each country's trade with Scotland. Let us emphasize the insight that this last aspect represents. It is easy to imagine someone of lesser insight succumbing to "America first" and ordering the bars alphabetically. The resulting figure would be much the worse (see figure 17.2).

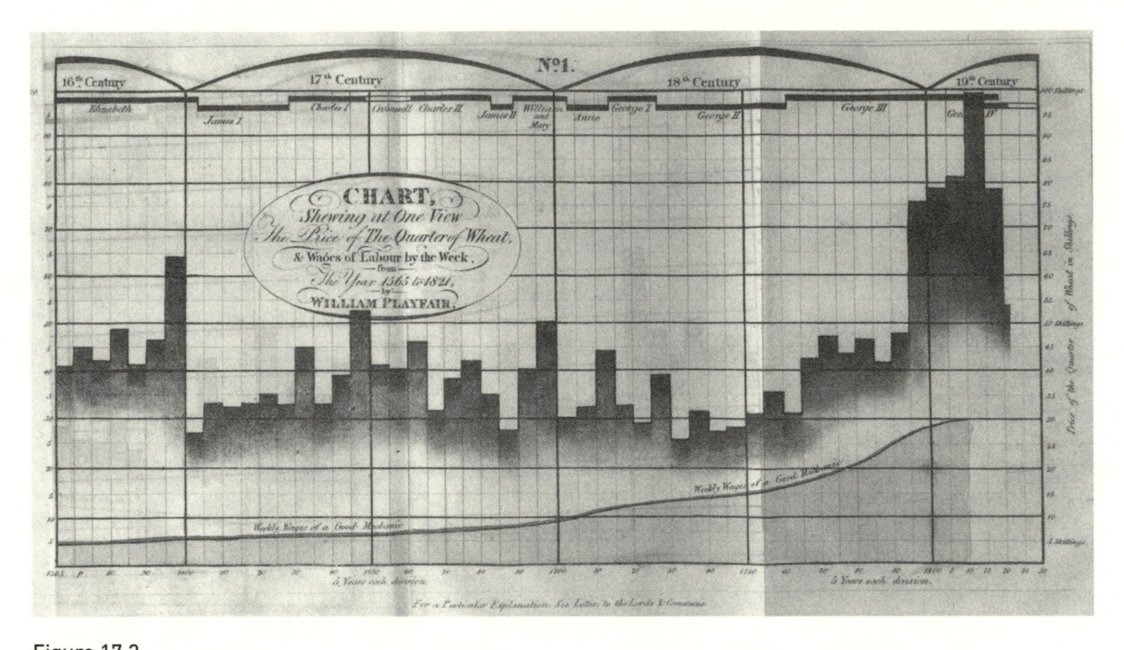


**Figure 17.3.**
A time series display showing three parallel time series: prices of a quarter of wheat (the histogram bars), wages of a good mechanic (the line beneath it), and the reigns of English monarchs from Elizabeth I to George IV (1565–1820). Our gratitude for this copy of Playfair's figure to Stephen Ferguson and the Department of Rare Books and Special Collections, Princeton University Library. (For the color version, see page C1.)

Playfair invented so many graphical forms, and because his taste was almost always impeccable, it is shocking to find one that has serious flaws. In his *Letter on our Agricultural Distresses, Their Causes and Remedies; Accompanied with Tables and Copper-Plate Charts Shewing and Comparing the Prices of Wheat, Bread and Labour, from 1565 to 1721*, he produced an apparently breathtaking figure (figure 17.3) showing three parallel time series: the price of a quarter of wheat, the weekly wages of a "good mechanic," and the name of the reigning monarch during each of the time periods shown.

His use of the line graph for such time series data was natural given how successful he had been in the past with similar data. But the inferences he wished to draw from this data set were subtly different from the ones he had made before. Previously the questions posed were "How have England's exports to *X* changed over the past 200 years?" Or "what has been the character of England's debt over the past 200 years?" Sometimes, when a second parallel data set was also included on the plot a new, but parallel, question pertaining to it was posed "How have England's imports from *X* changed over the past 200

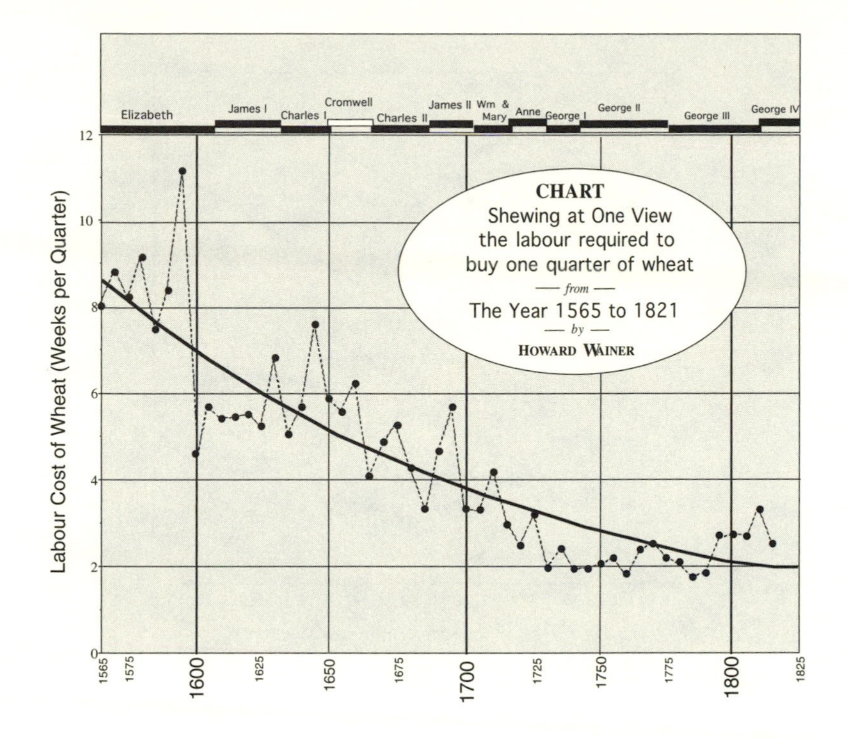


**Figure 17.4.** The times series from figure 17.3 recalculated to show the number of weeks of work required to buy one quarter of wheat. The individual data points are connected by a dotted line, a fitted quadratic is drawn through them (solid line).

years?" And then natural comparisons between the two data series were made —"When were exports greater than imports?"

All of these questions could be well answered by the format Playfair chose. But the data in figure 17.3 were meant to illustrate a deeper question. Playfair wrote (pp. 29–30):

> You have before you, my Lords and Gentlemen, a chart of the prices of wheat for 250 years, made from official returns; on the same plate I have traced a line, representing, as nearly as I can, the wages of good mechanics, such as smiths, masons, and carpenters, in order to compare the proportion between them and the price of wheat at every different period . . . the main fact deserving of consideration is, that *never at any former period was wheat so cheap, in proportion to mechanical labour as it is at the present time.* (Emphasis mine.)

Is this conclusion true? It is not easy to see in Playfair's figure as he produced it. Apparently Playfair was not fully acquainted with the

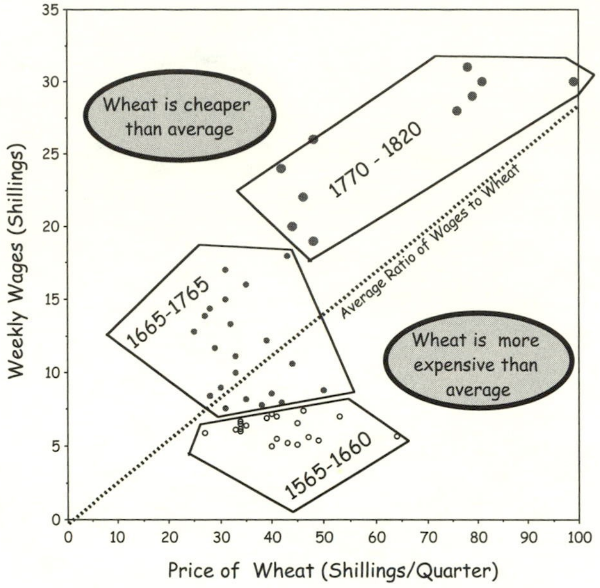

**Figure 17.5.**
A scatter plot of wheat price (on the horizontal axis) versus wages (on the vertical axis). The diagonal line represents the average ratio between the two over the 255 years depicted. Data points for three different eras are indicated.

benefits of combining variables into new variables to examine specific issues. In this instance, if we make a line plot in which the variable being displayed is the ratio "Labor cost of wheat (weeks/quarter)" the truth of Playfair's conclusion is evident.

Another way to look at the relationship between the cost of wheat and the amount of wages directly might be a scatter plot. But to construct a scatter plot that provides such an insight (see figure 17.5) takes a lot of work, and some experience in peering at them. Playfair might have considered a scatter plot, and rejected it as worthless—or at least as worthless for the sorts of time series data that he was primarily concerned with. Of course, had Playfair been interested in the relationship between parents' and children's heights he might have scooped Galton. But he wasn't and he didn't. Nobody's perfect.

# 18

## When Form Violates Function

There are a large number of graphical formats, each conceived to fit a particular purpose. Some formats, like scatter plots, are enormously robust and can provide useful insights under greatly varying circumstances. Others, like the pie chart, have a much more limited use. In figure 18.1 is an example of a pie chart format pushed beyond its limited range. Its components sum to 100% (the principal value of a pie chart), but that's all they do.

Few will grieve for the indignity imposed on the venerable pie by this artist's imagination. Pie charts are so often misused that another is hardly noticed. Moreover, pie charts are so pedestrian and so limited that we hardly care; akin to serving Thunderbird wine over ice cubes with a straw.

But we are, properly, outraged if anyone has the temerity to serve Dom Perignon with a twist over ice. Yet this is analogous to what happened in an article by Tom Stein in the November 2003 issue of *Wired*. In an effort to show how the 2000 merger of Time Warner and America Online (AOL) was a corporate disaster Mr. Stein decided to plot three variables (earnings, stock price, and AOL subscribers) as a function of time. For reasons that defy rational thinking he felt that this data set was admirably suited to be plotted as a flow map à la Minard's masterful depiction of Napoleon's Russian campaign (figure 18.2, for those who need to be reminded[1]). The width of Minard's river depicts both the size and location of Napoleon's army during its ill-fated inva-

**Figure 18.1.** Cartoon from *New Yorker*, October 13, 2003, p. 69. (© The New Yorker Collection 2003 Roz Chast from cartoonbank.com. All Rights Reserved.)

sion of Russia. Connected to the returning army is a plot showing the fatal cold during the Russian winter. The pairing of the rush of over 400,000 men crossing from Poland into Russia with the trickle of the 10,000 men returning makes vivid the tragedy of this march.

**Minard**

Charles Joseph Minard (1781–1870) had several careers. He was first a civil engineer and then an instructor at the École Nationale des Ponts et Chaussées. He later was an Inspector General of the Council des Pont et Chaussées, but his lasting fame derived from his development of thematic maps in which he overlaid statistical information on a geographic background. The originality, quality, and quantity of this work led some to call him "the Playfair of France."[2]

*(Continued)*

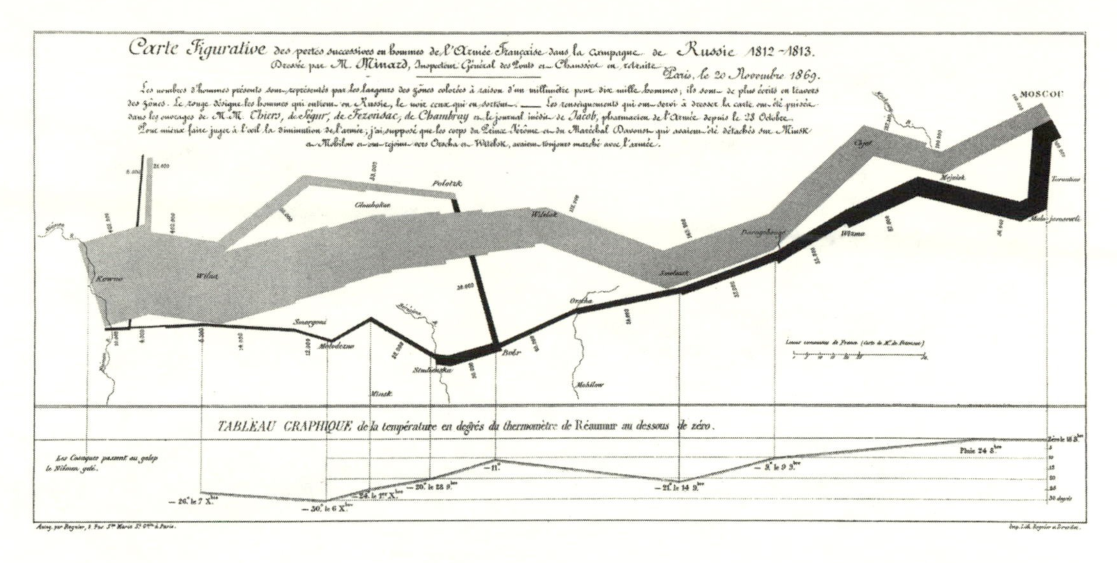


Figure 18.2.
Charles Joseph Minard's famous plot of Napoleon's ill-fated Russian campaign of 1812 (reprinted courtesy of Graphics Press, Cheshire CT). (See page C5 for a color version.)

(*Continued*)

His intellectual leadership led to the publication of a series of graphic reports, by the Bureau de la Statistique Graphique of France's Ministry of Public Works. The series (*l'Album de Statistique Graphique*) continued annually from 1879 until 1899 and contained important data on commerce that the Ministry was responsible for gathering. In 1846 he developed a graphical metaphor of a river, whose width was proportional to the amount of materials being depicted (e.g., freight, immigrants), flowing from one geographic region to another. He used this almost exclusively to portray the transport of goods by water or land. When he was eighty-eight years old he employed this metaphor to perfection in his 1869 graphic which, through the substitution of soldiers for merchandise, he was able to show the catastrophic loss of life in Napoleon's ill-fated Russian campaign. The rushing river of 422,000 men that crossed into Russia when compared with the returning trickle of 10,000 "seemed to defy the pen of the historian by its brutal eloquence." He included six dimensions in this plot: the location of Napoleon's army (its longitude and latitude), its direction of travel, and its size; and the dark line representing the retreating army is linked to both dates and temperature. This now-famous display has been called[3] "the best graph ever produced."

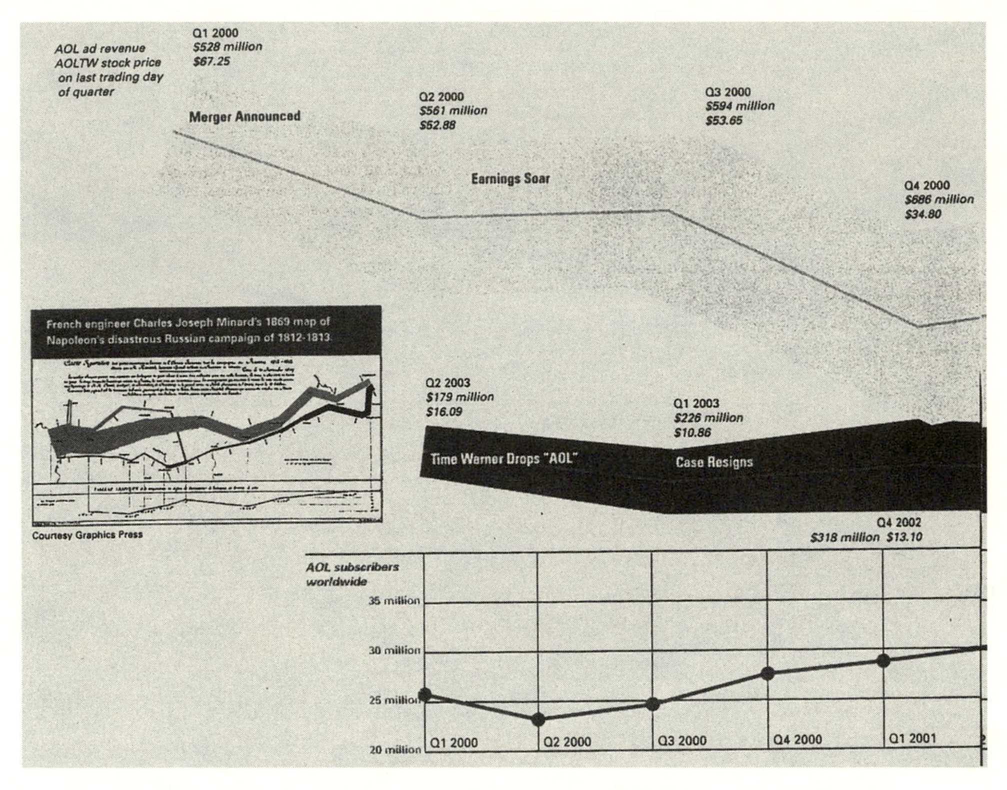


**Figure 18.3.**
AOL's death march. Produced by Tom Stein in *Wired*, November 2003, pp. 62–63. (See pages C8–C9 for a color version.)

What relationship does the plight of AOL–Time Warner have to Minard's six-dimensional masterpiece? Let us consider Stein's display shown as figure 18.3. The river of AOL–Time Warner sweeps across two pages changing from pale to dark at the end of the fourth quarter of 2001 when it reverses its field and heads west. What does the river represent? Other than as a general metaphor for AOL–Time Warner's declining fortunes, it is hard to figure out. The width of the river ends up being about half what it was at the time plans of the merger were announced. Advertising revenues dropped from $528 million to $179 million, a decline of two-thirds, so that can't be what is being represented. The stock price dropped from $67.25 to $16.09, a decline of three-quarters. So that's not it either. What does the width of the river represent? And what does the location of the river on the page mean? As the river drops sharply after the third quarter of 2000, advertising revenues increased from $594 million to $686 million. And why does

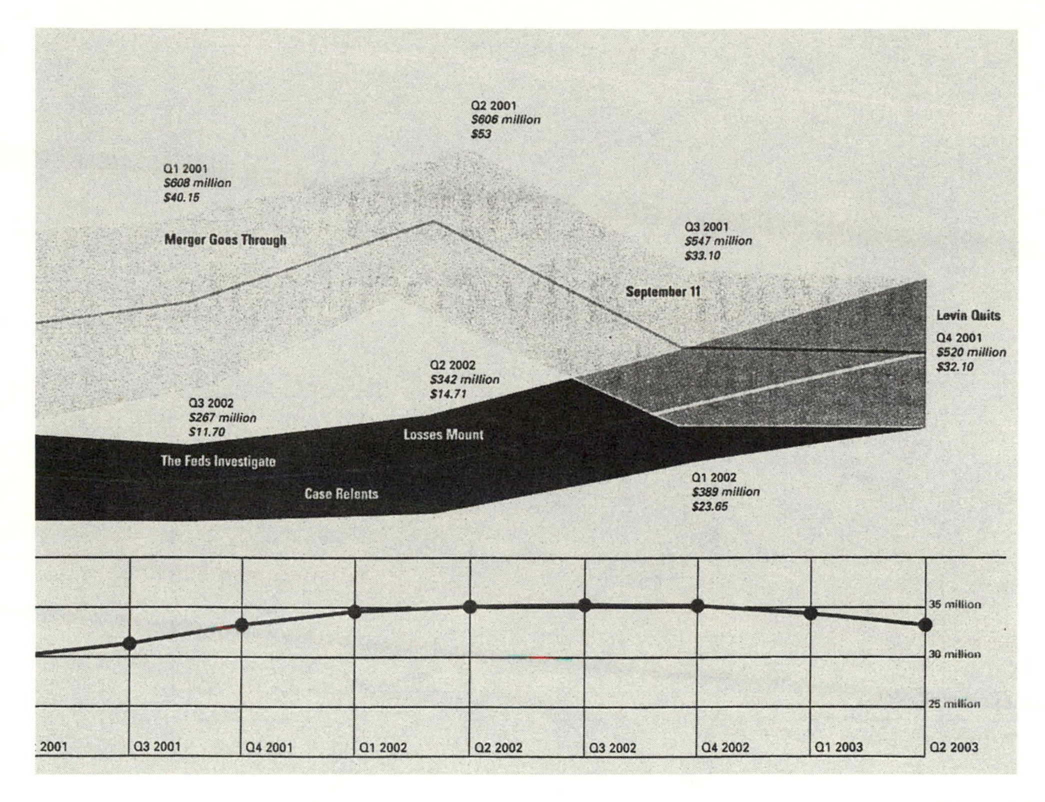


Figure 18.3.
*(Continued)*

it reverse itself when chairman Gerald Levin announced his retirement on December 5, 2001? The only conclusion that I can reach was that the graphist who drew this was trying to mimic Minard's form without regard to its content.

What other graphical depiction might serve better? Briefly, almost any other, but we are offered a plausible hint from the one aspect of Stein's display that is effective—the line graph of AOL subscribers at the bottom. This plot unambiguously shows us that, although the fortunes of AOL–Time Warner may have been in steep decline, the subscriber list did not begin to shrink until the end of 2002. Perhaps using this same format for both advertising revenues and stock price would yield an informative result. Such a display is shown in figure 18.4.

Four aspects of the data, which were previously veiled, are now clear.

(1) The actual merger boosted the stock price, but not enough to make up for the drop suffered when it was first announced.

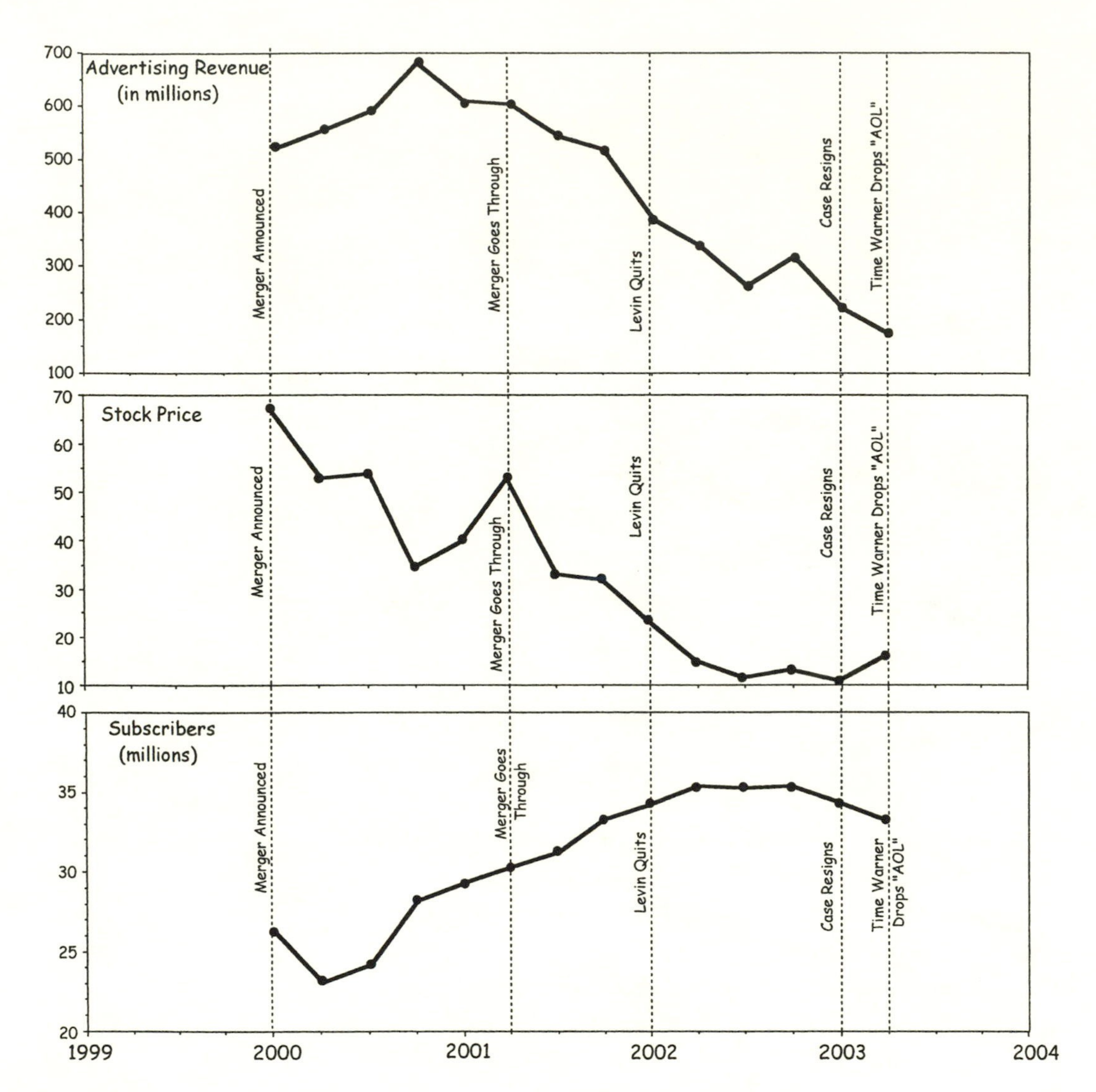


**Figure 18.4.**
Alternative formulation of AOL–Time Warner data.

(2) When the merger was finalized the decline in advertising revenue continued unabated.
(3) Gerald Levin's departure changed nothing, although when AOL–Time Warner Chairman Steve Case resigned the stock price did show a bump.
(4) The curious imperviousness of both stock price and advertising revenue to the continued growth of the subscriber base (until the end of 2002).

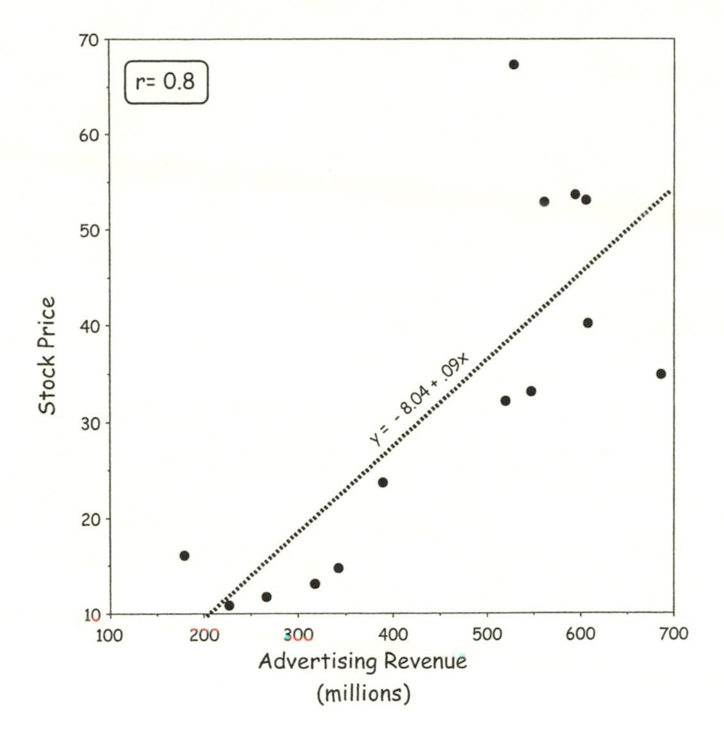


Figure 18.5.
Scatter plot of AOL–Time Warner's advertising revenue and its stock price quarterly from the beginning of 2000 until the end of 2003.

Of course, some characteristics of the data would be revealed more fully with other formats. For example, if we wish to look at the extent of the covariation between advertising revenue and stock price a scatter plot is very helpful (see figure 18.5) and indicates that, on average over this time period, for every $100 million of increased advertising revenue the stock price climbed nine dollars, or more precisely,

$$\text{Stock Price} = .09 \times \text{Ad Revenue (in millions)} - 8.04.$$

But I may be too harsh in my criticism. Perhaps Stein's use of Minard's metaphor is more subtle. Maybe his point was that bad corporate decisions are well represented by bad graphical decisions; roughly akin to using poor handwriting to recount an ugly story. Such metaphoric parallelisms are rarely a formula for effective communication.

# 19

## A Graphical Legacy of Charles Joseph Minard: Two Jewels from the Past

The evil that men do lives after them;
The good is oft interred with their bones;

—Shakespeare, *Julius Caesar*,
Act 3, Scene 2

In chapter 18 we met Charles Joseph Minard (1781–1870) the French civil engineer turned economic geographer-cartographer whose legacy violates Marc Antony's funeral oration for Julius Caesar. While Minard is known today principally because of his graph of Napoleon's march on Moscow (figure 18.2 has been reproduced many times as well as becoming a wonderful wall poster[1]) that was published in the year before his death, his graphical work was much broader than that one plot. Napoleon's Russian campaign is not Minard's only graphical legacy.

Cartographers call the format of the Napoleon plot a flow map, which was not original with Minard. Henry Drury Harness produced thematic maps in 1837 that showed the flow of people and freight in Ireland. But Harness's maps seem to have had little impact, perhaps because, as worthy a topic as rail traffic is, it does not evoke the same feelings as a chart showing the death of 400,000 men. In graphics, both form and content matter; a lesson that Minard apparently did not grasp immediately either.

Minard began his study of civil engineering at the Ecole Polytechnique when he was sixteen. He further studied at the Ecole Nationale des Ponts et Chaussées, a school that trained the engineers responsible for building and maintaining France's roads, canals, railways, and port facilities. He eventually became superintendent of that school and later rose to the rank of inspecteur général. Over the course of his career his interest focused on economic geography, especially those aspects of the subject that dealt with the movement of goods. Over the last twenty-five years of his life he produced more than fifty maps, most of which were flow maps. At the very end he used all he had learned of the techniques of flow maps to produce his final masterpiece, the chart of Napoleon's march. This plot was paired in publication with a parallel, but lesser known, flow map tracing the path of Hannibal's army from the beginning of its campaign in Spain with 96,000 troops to its end in Italy with only 26,000. Apparently crossing the Alps with elephants wasn't a whole lot smarter than crossing the Russian Steppes in winter.

Although Minard's economic flow maps did not achieve the same level of popularity in the twenty-first century as his maps depicting the dangers of war, they were enormously influential. Victor Chevallier, the Inspector General of French bridges and roads, in his extensive obituary for Minard, reported that in the decade from 1850 to 1860 all of the Ministers of Public Works in France had their portraits painted with one of Minard's data displays in the background.*

Of greater longer-term value was the influence Minard's work had on the practice of statistical communication. For twenty-one years, from 1879 until 1899 the Bureau de la Statistique Graphique of the Ministry of Public Works annually published a large-format chart book. This series, *l'Album de Statistique Graphique*, rivals in sophistication and polish the very best statistical atlases published today. And most of the displays found within it derive very directly from earlier versions designed by Minard. Each volume of the Album consists of a few pages of explanatory and interpretative text and at least twenty

*See the "NEW" section of http://www.edwardtufte.com for a reproduction of Minard's map of Hannibal's march as well as an English translation of Chevallier's obituary of Minard.

graphical displays. The displays are very large, full-color, foldout affairs, extending to three or four times the size of the Album.

The productions of these Albums were a perfect manifestation of the zeitgeist. In 1878 the French physiologist Etienne Marey, expressed the value of graphical representation. His graphic schedule of all the trains between Paris and Lyons provides a powerful illustration of the breadth of value of this approach. And, on the off chance that someone might have missed the point he provided an explicit conclusion,

> There is no doubt that graphical expression will soon replace all others whenever one has at hand a movement or change of state—in a word, any phenomenon. Born before science, language is often inappropriate to express exact measures or definite relations.

Marey was also giving voice to the movement away from the sorts of subjectivity that had characterized prior science in support of the more modern drive toward objectivity. Although some cried out for the "insights of dialectic," "the power of arguments," "the insinuations of elegance," and the "flowers of language," their protestations were lost on Marey, who dreamed of a wordless science that spoke instead in high-speed photographs and mechanically generated curves; in images that were, as he put it, in the "language of the phenomena themselves."

Historians have pointed out that "Let nature speak for itself" was the watchword of the new brand of scientific objectivity that emerged at the end of the nineteenth century. In a 1992 article, Lorraine Daston and Peter Galison pointed out that "at issue was not only accuracy but morality as well: the all-too-human scientists must, as a matter of duty, restrain themselves from imposing their hopes, expectations, generalizations, aesthetics, and even their ordinary language on the image of nature." Mechanically produced graphic images would take over when human discipline failed. Minard, Marey, and their contemporaries turned to mechanically produced images to eliminate human intervention between nature and representation. "They enlisted polygraphs, photographs, and a host of other devices in a near-fanatical effort to produce atlases—the bibles of the observational sciences"—documenting birds, fossils, human bodies, elementary particles, flowers, and economic and social trends that were certified free of human interference.

The problem for nineteenth-century atlas makers was not a mismatch between world and mind, as it had been for seventeenth-century epistemologists, but rather a struggle with inward temptation. The moral remedies sought were those of self-restraint: images mechanically reproduced and published warts and all; texts so laconic that they threatened to disappear entirely. Seventeenth century epistemology aspired to the viewpoint of angels; nineteenth century objectivity aspired to the self-discipline of saints. The precise observations and measurements of nineteenth century science required taut concentration endlessly repeated. It was a vision of scientific work that glorifies the plodding reliability of the bourgeois rather than the moody brilliance of the genius.

What follows are two jewels, plucked, not quite at random, from the 1892 Album to provide a small illustration of the impressive state of graphic art in nineteenth-century France.

## JEWEL 1: TRAFFIC AND PROFIT FOR FRENCH RAILWAYS IN 1891

Figure 19.1 is a flow map, obviously a lineal descendent of Minard's plot of Napoleon's march. Like its distinguished forebear, this display also integrates multivariate complexity "so gently that viewers are hardly aware that they are looking into a world of six dimensions."[2] This display format has often been repeated and is so termed a "cornerstone" display. The lines on the map show the location of each of the rail lines. The thickness (on a scale of 1 mm per 20,000 francs) shows the per kilometer profit of each link. A blue line indicates a loss, with the width depicting the size of the loss. The text surrounding this display points out that 1891 showed growth in gross income over the prior year, but a decline in net income due to an increase in operating expenses (retirement pensions, salary increases, reduction in working hours). It also refers the reader to other displays that capture different aspects of the railways as well as parallel displays of income from waterways. The legend in the lower left corner provides a numerical summary broken down by the various railroad companies. The power of the display is in how well it serves multiple purposes. A quick glance shows the unsurprising result that the greatest profits lie where there is the greatest traffic—to and from Paris, Lyons, and the ports. More careful study provides details about specific railways. And using this

Figure 19.1. The net profits per kilometer of French railways in 1891. (See the color version on pages C10–C11.)

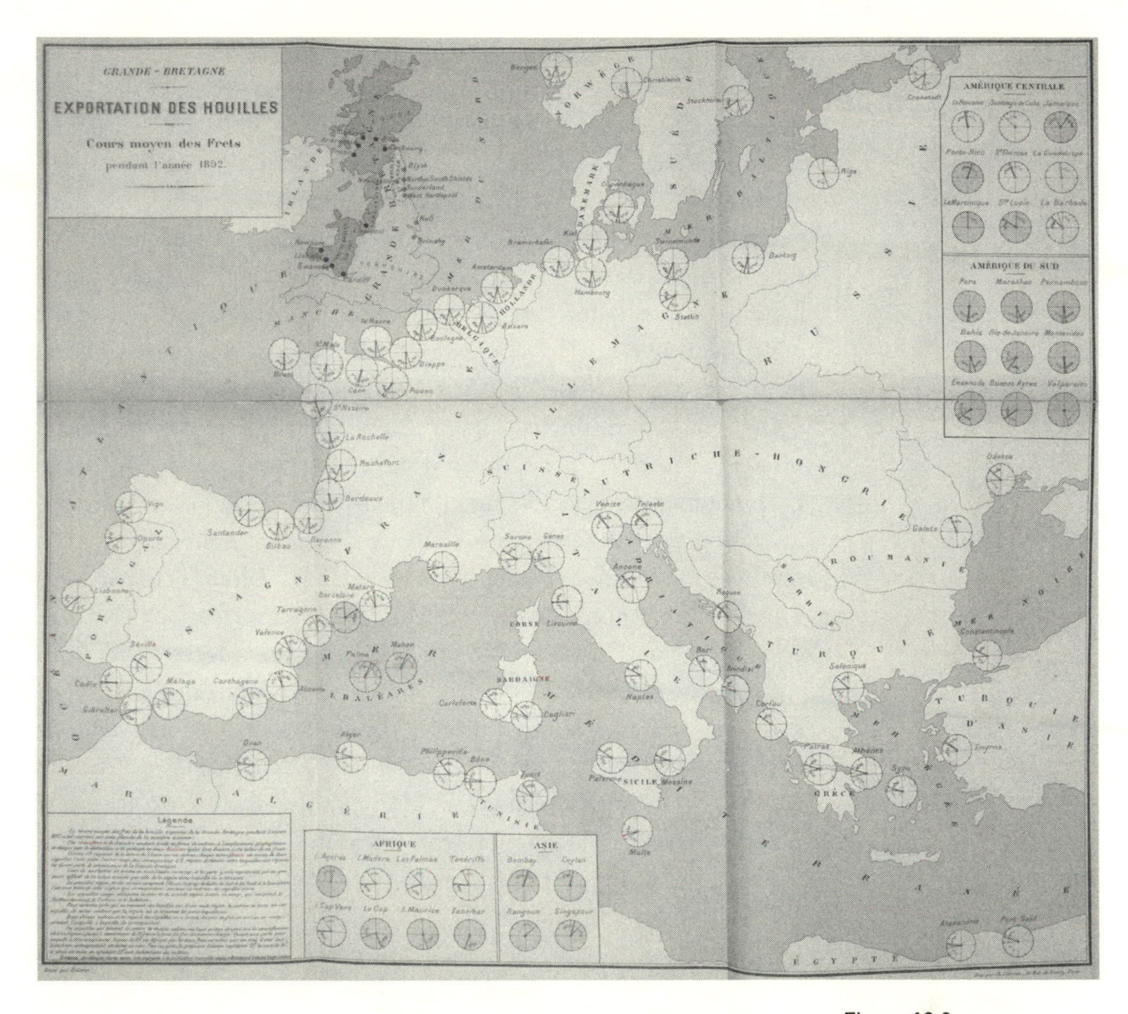


Figure 19.2.
Average price for transportation of English coal in 1892. (See color version on pages C12–C13.)

display in conjunction with others allows one to assess the efficacy of various pricing policies. Indeed, 1891 marked the beginning of a new pricing policy that yielded big increases in traffic, but, unfortunately, bigger increases in costs.

## JEWEL 2: WHERE ENGLISH COAL GOES AND HOW MUCH IT COSTS TO GET IT THERE

Figure 19.2 is another flow map, but it uses a different graphical metaphor to depict the movement of coal from one of two English mining regions to a variety of destinations. The British ports were divided into two groups. In the original multicolored version (pages C12–C13), the first, colored black, consisted of Southern and Northern Wales, Lancashire, and Scotland; the second, colored red, contained Yorkshire, Northumberland, and Durham. Because the shipping cost of coal generally varied between 3 and 12 francs per ton, the metaphor of a clock face to represent it must have suggested itself naturally. Each destination is marked by a clock face with one or two hands. The color of the hand corresponds to the region of origin, and the number that the hand points to tells the shipping cost. Most destinations received coal from both regions and so had two hands. It is easy to see that nearby destinations on the English Channel, the North Sea, and the Baltic all received coal from both sources at prices that were in the 5–7 franc range. But apparently crossing into the Mediterranean was expensive, increasing the cost to 9–12 francs per ton. The metaphor is modified for those few locations where the price exceeds 12 francs per ton (e.g., some Mediterranean islands). When this happens the clock faces are colored pink and range from 13 to 24 francs. This suggests that both the shipping distance and the size of the market determine the cost of shipping coal. This interpretation receives further support when we note that although Stockholm and Riga are not appreciably closer to British supply than Gibraltar and Malaga, the shipping cost is less. It is not far fetched to assume that this reflects the fact that the northern ports are supplying better customers. How much coal was needed in nineteenth-century Algiers?

This display too can serve many purposes. Principally, it can help study the conditions in which competition occurs: between maritime navigation and river navigation and railroads; and also between French domestic coal production and that of the English coal regions.

As it stands this display carries five dimensions (the location—longitude and latitude—of coal destination, the origin of the coal, and the cost of the coal for each of two origins). It would have been easy to carry one more, by letting the size of the clock face be proportional to the total amount imported. But it would be graceless to carp about what is missing, when what is there is done so beautifully.

The graphical heritage of Charles Joseph Minard lives on.

20

# La Diffusion de Quelques Idées: A Master's Voice

Jacques Bertin (1918–) is a French semiologist, trained in Paris, whose seminal work *La Semiologie Graphique* (1969) laid the groundwork for modern research in graphics. Forty years after its publication it still provides important lessons to all of those interested in the effective display of quantitative information. Until his retirement in 1984 he was the director of the Laboratoire Graphique in the Ecole des Hautes Etudes en Sciences Sociales in Paris. Bertin, together with John Tukey and Edward Tufte, form a triumvirate that became the key twentieth-century contributors to the understanding of graphical display.

In 2003 I got a note from Bertin along with a twenty-page document with his byline, dated 2002, and entitled (not surprisingly) "La Graphique." A footnote proclaims that it contains extracts from *La Semiologie* and his shorter work, *La Graphique et le Traitement Graphique de l'Information*, but that some aspects are new.

His cover note said (in its entirety),

> A Howard Wainer
> à qui je dois la diffusion de quelques idées, aux USA.
> Bien amicalement,
> J. Bertin*

* A rough translation is, "To Howard Wainer, to whom I owe the diffusion of some ideas to the USA. Best Wishes, J. Bertin."

My arithmetic tells me that Bertin was eighty-four years old when he wrote this. While I remain in awe of the intellectual vitality that this mailing suggests, I also feel that if I am to aid in the diffusion of these new ideas I ought not tarry too long.

There are a number of ideas within the twenty pages worthy of discussion, although many of them are contained in the aforementioned publications. I will focus on just one (see figure 20.1) taken from section 6 of this document.

In the first panel we see a map of the United States. In each state big enough to contain it, is a list of those raw materials that state produces for the chemical industry.

How should these data be displayed? Obviously, the answer to this question depends strongly on what questions the data on the map are meant to answer. There are, according to Bertin, essentially two basic questions:

(1) Which raw materials does this state produce?
(2) Which states produce this particular raw material?

The answer to each question requires a different graphical construction. Bertin analyzes this problem by observing that there are two dimensions of the data, the states ($X$) and the raw materials ($Y$). Thus the questions are which $X$'s have $Y$ or which $Y$ are produced in $X$. The graphical problem is that the two dimensions of the plane are used up with geography and so the various materials must be represented some other way than spatially.

Bertin begins the analysis with the construction of a graphical matrix (figure 20.2) only hinted at in the summary in the center of figure 20.1. The rows and columns of this matrix are then permuted to yield some sort of simple structure in which, to whatever extent possible, similar states are placed adjacent to one another, as are similar raw materials. This rearrangement aids in the formation of subsequent plots.

The original map or the original arrangement of the raw data matrix, in which states are arranged alphabetically, allows an easy answer to the first question, and so Bertin directs his attention to the second

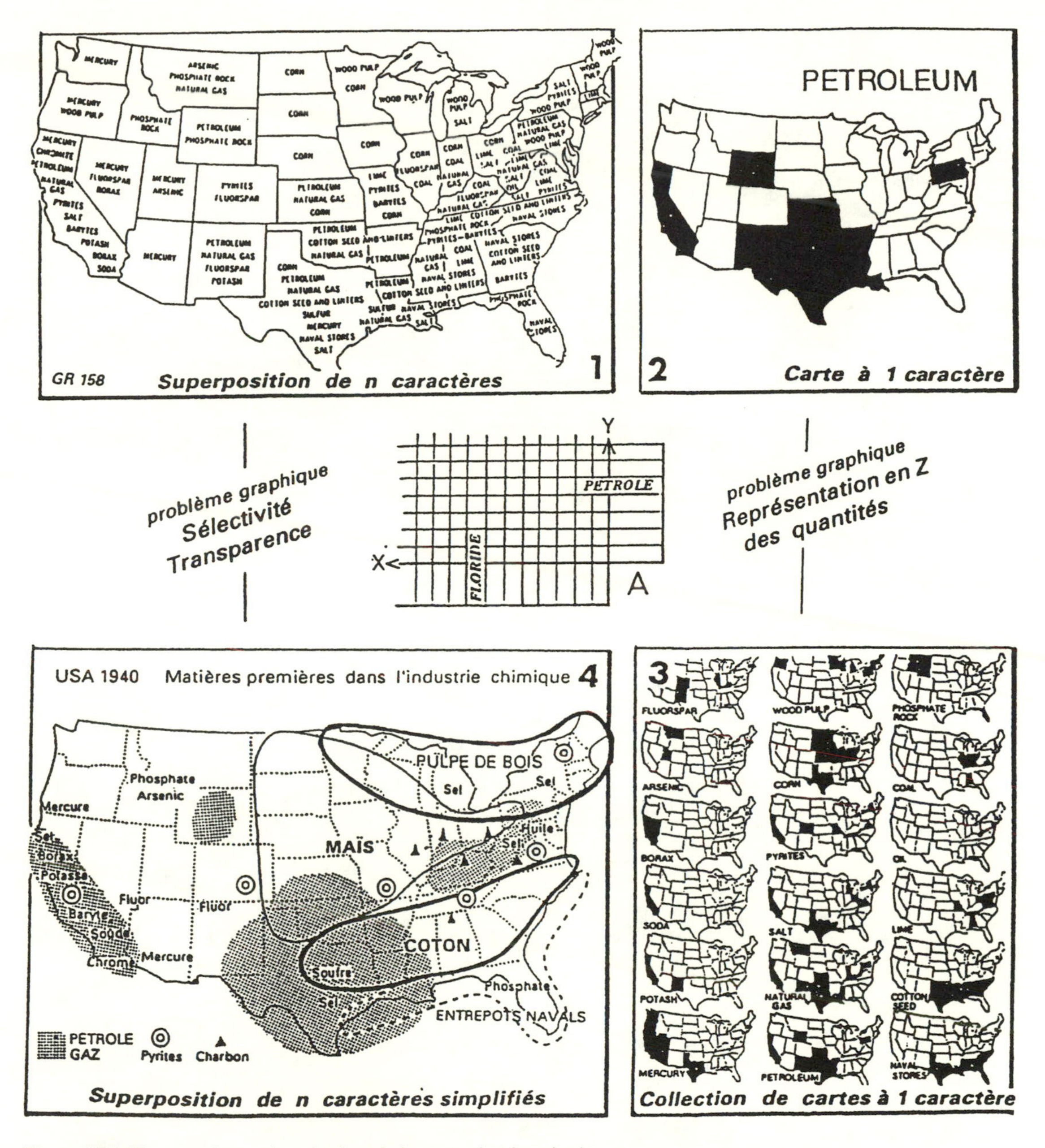


Figure 20.1. The transformation of a data-laden map that is suited to answer one question into a form suited to answer another.

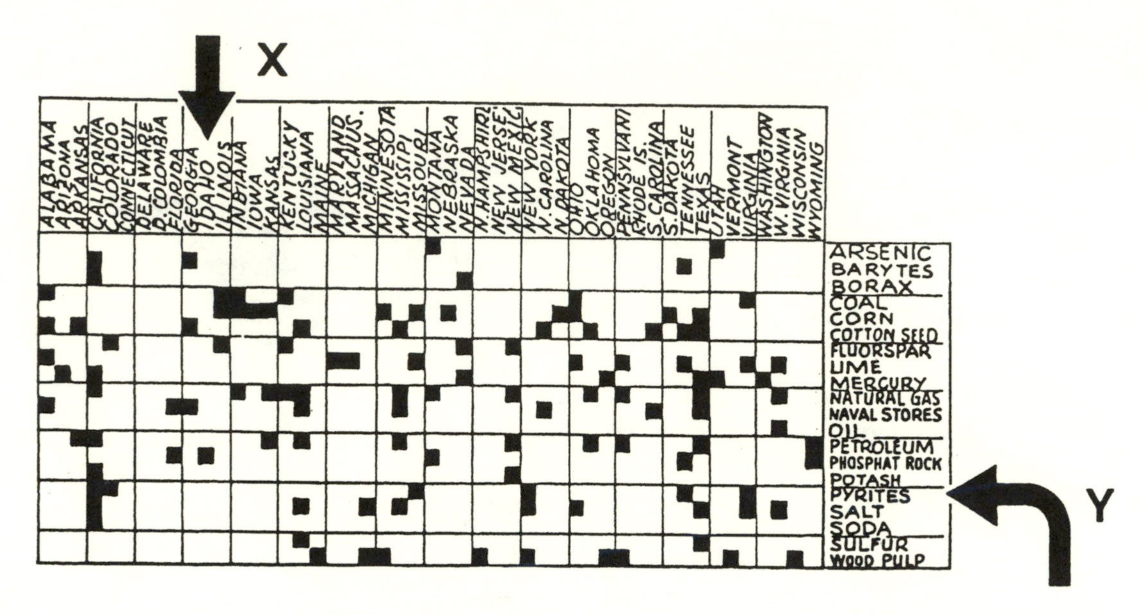


**Figure 20.2.** The graphical matrix that contains the content of the map in a form suitable for reorganization on the basis of the data, not the geography.

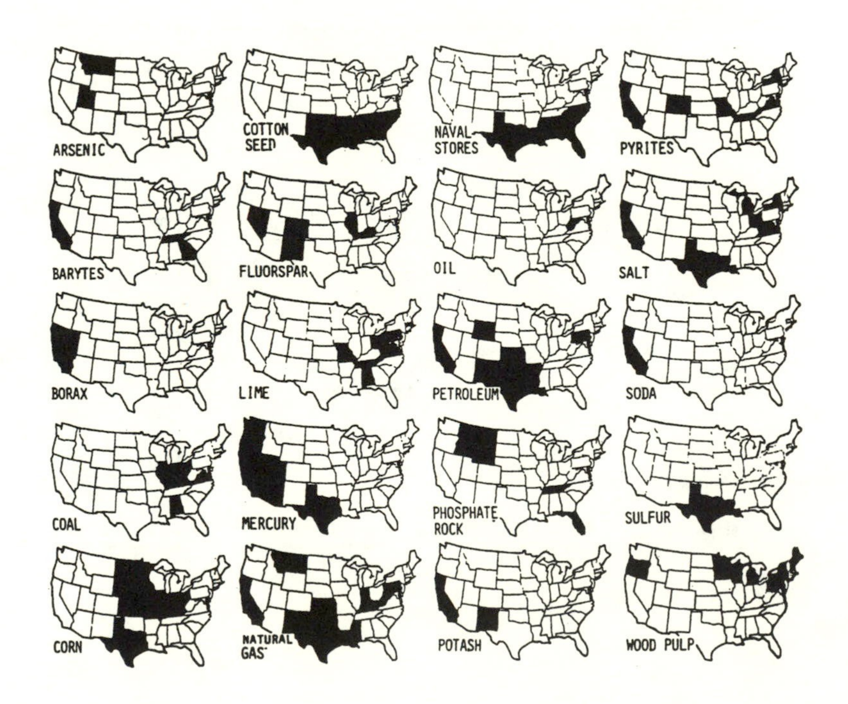


**Figure 20.3.** A collection of one-variable maps arranged alphabetically.

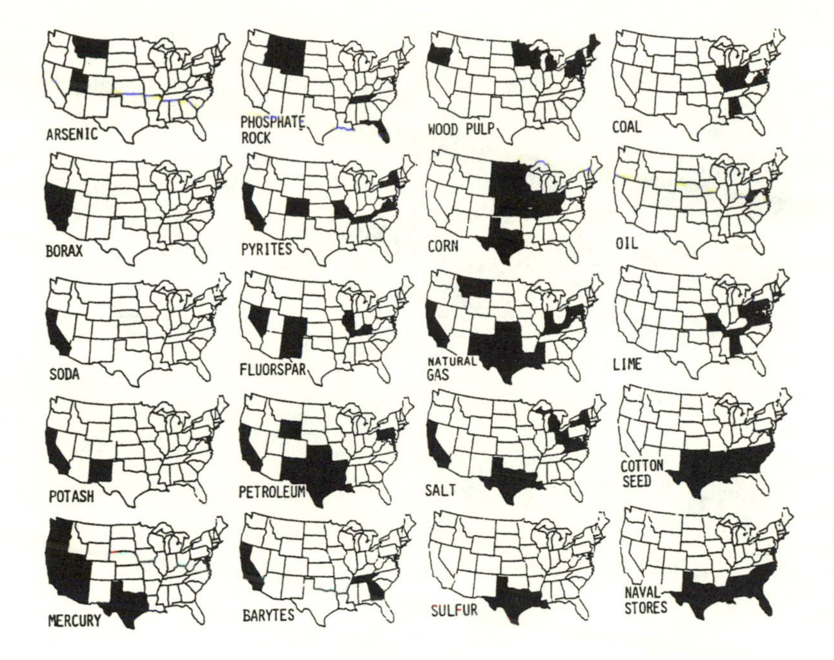


**Figure 20.4.**
A collection of one-variable maps arranged geographically.

question. From the original construction this question cannot be answered without reading every word on the map and remembering what each state produces in order to get the correct answer. This is pretty close to impossible.

Panel 2 of figure 20.1 is an example of a plot that answers question (2) for one product, petroleum. It is a simple matter to prepare similar maps for each product (panel (3)). These separate maps can be then be arranged in any way that is convenient. They can be alphabetical to make locating any particular raw material easy (figure 20.3) or they can be arranged geographically (figure 20.4) to enable us to uncover regional traits.

Bertin also shows (panel 4 of figure 20.1) how everything can be fit onto a single map, at the simplifying cost of omission of some minor raw materials.

Earlier in his document Bertin offers alternative names for some well known graphical formats (figure 20.5) in which pie charts are

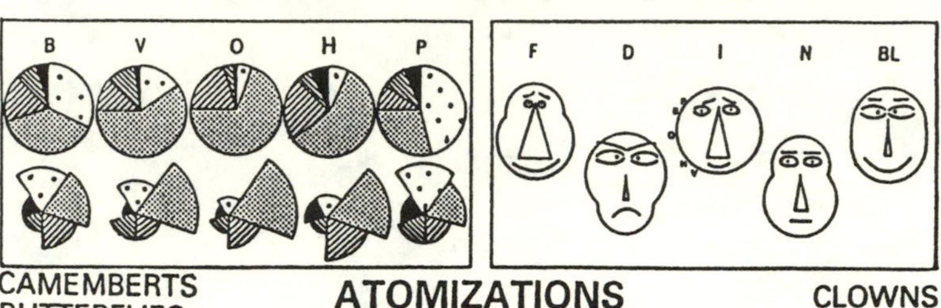


Figure 20.5.
Three well-known graphic icons with new names.

renamed "Camemberts," Florence Nightingale's *Roses* are dubbed "Butterflies," and Herman Chernoff's *Faces* are translated to "clowns." I suspect that butterflies and clowns are unlikely to catch on, but from now on I, for one, will always apply the new term to pie charts—they have always seemed pretty cheesy to me.

# 21

# Numbers and the Remembrance of Things Past[1]

A single death is a tragedy; a million deaths is a statistic.

—Joseph Stalin (1879–1953)

## 21.1. INTRODUCTION

Unquestionably cold and cruel, Stalin's sentiment sadly captures an aspect of human psychology. The mind is limited in its capacity to fathom cataclysmic events. Great numbers of deaths, particularly if they are distant in time or space, typically do not elicit the same reaction as fewer deaths nearer to us. Sponsors and designers of memorials face the challenge of stirring emotion, memory and understanding.

Bringing losses of great magnitude to a human scale is a critical function of memorials. Consider the enormous impact of the sweeping wall of the Vietnam Memorial, inscribed with the individual names of the more than 50,000 Americans who lost their lives in that conflict. More recently, it has been decided that the memorial for those who died when a hijacked jetliner crashed into the Pentagon will take the form of a garden with 187 benches, one for each of the victims.

Ordinarily, memorials are created by individuals who themselves are detached from the events they seek to transform through artistic imagination. Rarely do the living, in anticipation of the end, create a tangible monument to their demise. However, some victims of the Holocaust created precisely such memorials, which have been found and preserved. Juxtaposed against the records of the perpetrators, these chronicles of communities under siege convey both the tragedy and the pathos of the times. Some contain statistical tables and charts, usually thought of as dry and dispassionate, that carry with them not only the uncontestable authority of fact but also the capacity to elicit the viewer's emotions.

### 21.2. THE HOLOCAUST

While World War II began in September 1939, a critical point in Hitler's plans for the "Final Solution" was June 22, 1941. On that date, German armies invaded eastern Poland, which was then occupied by the Soviet Union. Their relentless advance soon brought them to the borders of the Soviet Union itself as well as of the Baltic States that were under its rule. As soon as control of a region was secured, the Nazis established ghettos for the Jewish population. Lithuania, with a large Jewish population, was no exception. Ghettos were formed in Vilnius, Kovno, and other cities.

During the course of the occupation of Lithuania the Nazis initiated a series of actions that resulted in the deaths of over 136,000 Jews. Initially, the Germans in elaborately illustrated reports meticulously documented these murders, but their production was reduced as the war continued. Indeed, in October of 1943 SS chief Heinrich Himmler, in a speech to his subordinates, rationalized the minimized record keeping, saying, "This is an unwritten and never-to-be-written page of glory in our history." Nonetheless, some documents were created. The map shown as figure 21.1, titled "Jewish executions carried out by *Einsatzgruppe A*," was used to illustrate a summary report by *Einsatzgruppe A* leader Walther Stahlecker, a Brigadier General in the SS. (Note that all page numbers in the captions of figures 21.1–21.7 refer to the location of the figure in *Hidden History of the Kovno Ghetto.*)

While the Nazis were tabulating their triumphs, others also kept records of these atrocities. In many ghettos, Jewish leaders organized committees to keep official chronicles of daily life. Indeed, from the

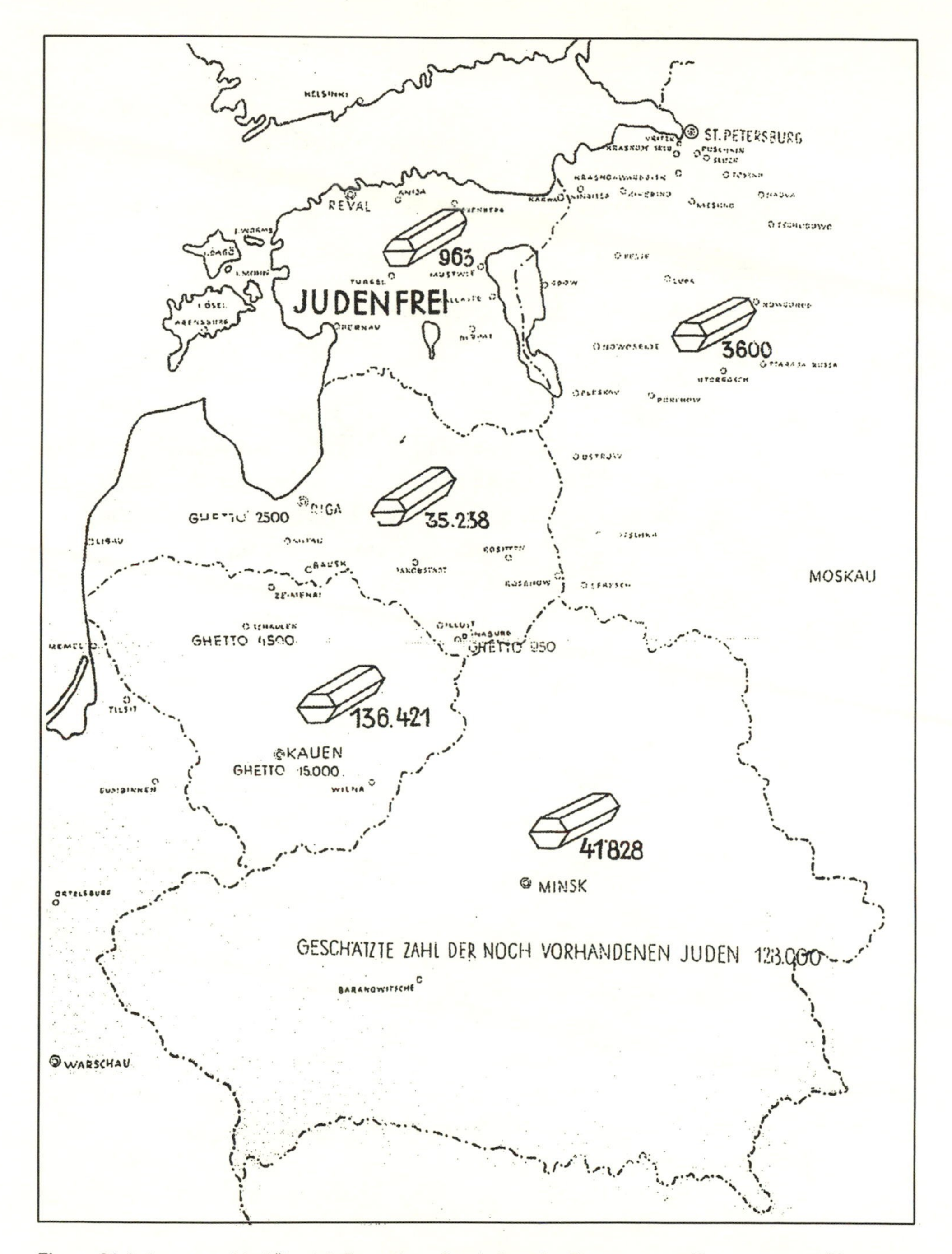


**Figure 21.1.** A map entitled "Jewish Executions Carried out by *Einsatzgruppe A*" was stamped "Secret Reich Matter," and was used to illustrate a summary report. The map records the number of Jews executed in the Baltic States and Belorussia following the German invasion. (Page 20.)

outset of the establishment of the Kovno Ghetto, Elkhanan Elkes, chairman of the Jewish council (the Ältestenrat), asked Kovno's Jews to write their own history as a legacy for future generations. The graphs and tables that were produced by the Kovno ghetto tell a powerful story.

### 21.3. SOME GRAPHIC MEMORIALS

The data displays that were produced within the Kovno ghetto varied in their sophistication and technical virtuosity. The ghetto's Statistics Office report "The September and November Census in 1941" is a simple plot showing the population losses due to the "Great Action" of October 28, 1941. The format is a familiar population pyramid with age running down the center, males to the left and females to the right. The total length of each bar represents the population in September, while the shaded portion of each bar depicts the number still alive in November. Its simplicity belies its chilling message.

Another group within the ghetto prepared an isotype plot* (figure 21.3) of these same data that showed, by gender and age, the victims of this "Great Action." Each tiny icon represents 200 victims.

Similar data showed up in a later report in a more polished format (figure 21.4), in which the bottom line shows the age distribution of the Ghetto population in November and the top line in September. The left and right panels divide the losses between men and women, respectively.

Sometimes the displays are more subtle. The table shown as figure 21.5 was compiled by the ghetto housing office in 1942. It shows street by street the number of dwellings, apartments, rooms, kitchens, square meters of space, population, and square meters per person. The number 2.95 square meters/person at the bottom right hand corner indicates the extent of the crowding. It doesn't take a rich imagination to visualize what life was like for the 16,489 inhabitants under such difficult living conditions.

Medical summaries were also represented graphically. Dr. Jacob Nochimowski prepared a report on May 9, 1943, on labor and illness

*Isotype graphs, made popular by Otto Neurath in the 1930s, show amounts by reproducing a single image proportional to that amount.

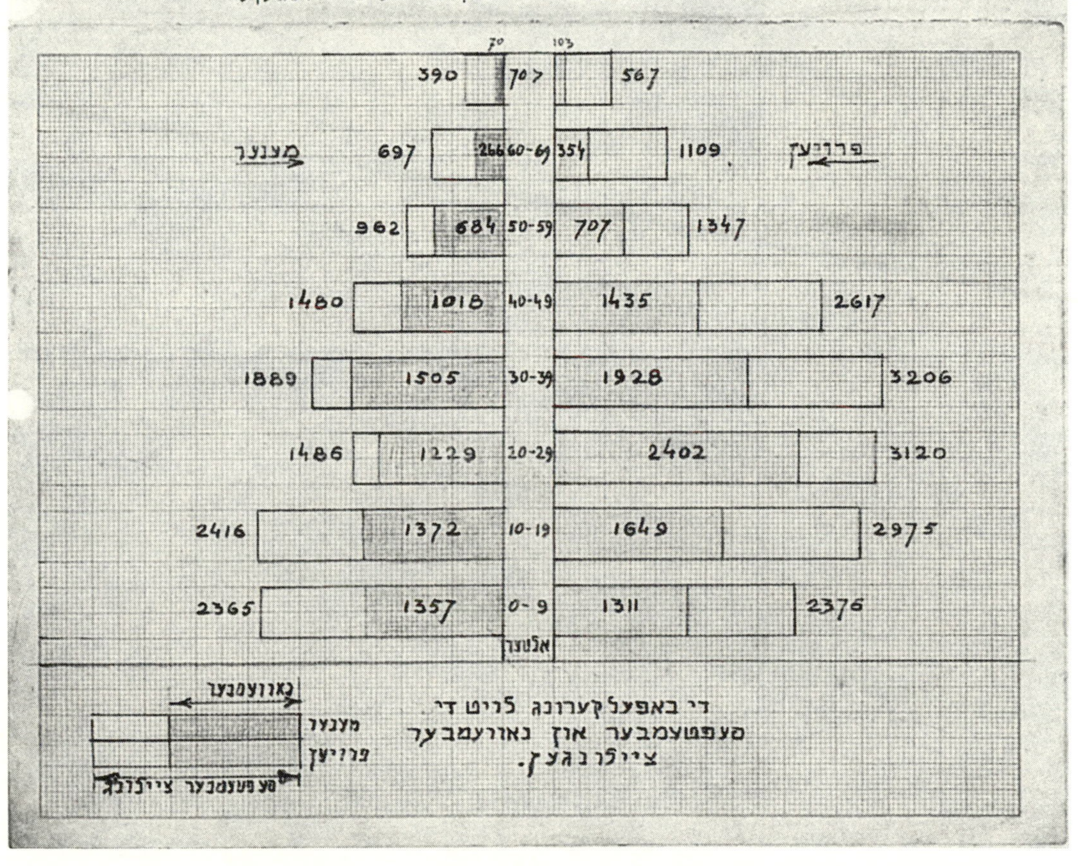

6

אין סעפטעמבער 148 אף 100 מענער און אין
נאוועמבער 132 " 100 " .

ווי ווייט די פארהעלטעניש העכסטע ווייכן אפ פון דער נארמע באווייזן די פארגלייכן מיט
אייראפיישע לענדער. די/פראפארץ צווישן מענער און פרויען האט באטראפן 114 פרויען אף
100 מענער (אין עסטלאנד און אין לעטלאנד). די קלענסטע צאל אין איראפע (אין האלאנד)
איז געווען 101 פרוי אף 100 מענער (1934) .

די ביילגנדע דיאגראמע גיט א גראפיש בילד פון ביידע ציילונגען.



Figure 21.2. The population losses in the Kovno Ghetto due mainly to the "Great Action" of October 28, 1941. Males are represented on the left, females on the right. The shaded portion represents those still surviving in November. The central column shows the age groups depicted. (Page 69.) (See color version on page C14.)

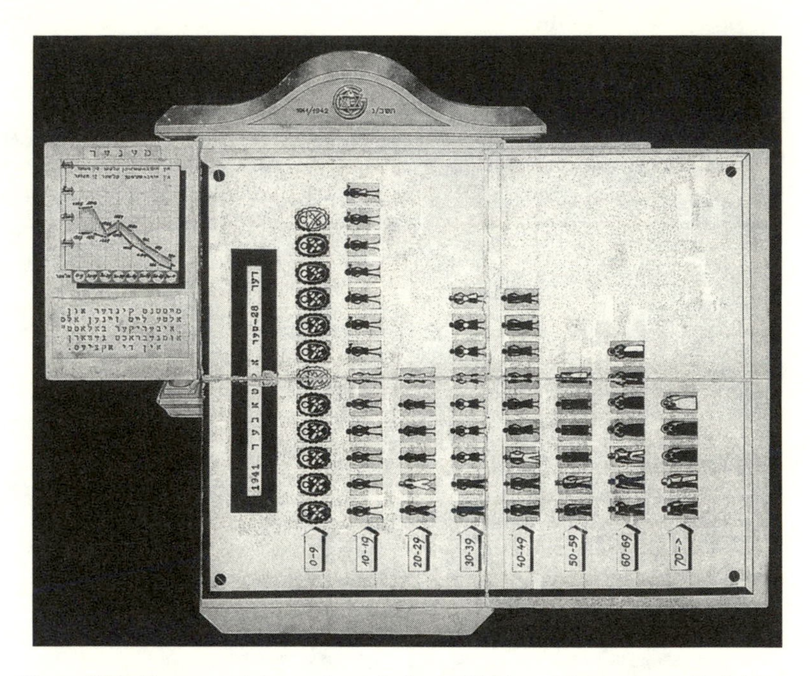


Figure 21.3. The same data as shown in figure 21.2 in a different form. Each icon represents about 200 of the victims. (Page 160.) (See color version on page C15.)

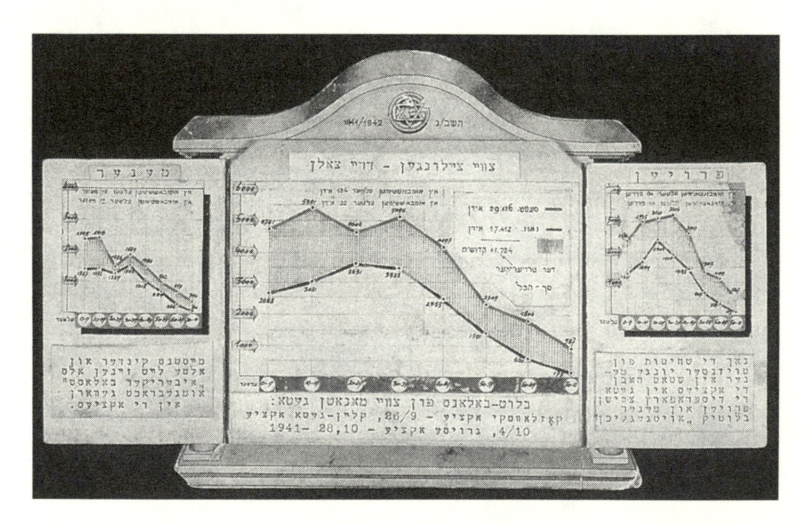


Figure 21.4. The same data as in figures 21.2 and 21.3, shown this time as line charts, both total (in the center) and separately for men (on the left) and women (on the right). The upper lines represent the age distribution of the ghetto population in September, the lower line in November. The shaded portion then represents those murdered between those two dates. (Page 158.) (See color version on page C16.)

Wohnungsamt
der
jüdischen
Ghetto-Gemeinde

Stand
des Ghettos
zum 31/XII 42

| | Ghetto-Gemeinde | Anzahl | | | | Fläche in m² | | Einwohnerzahl | Durchschnittsfläche pro Person in m² |
|---|---|---|---|---|---|---|---|---|---|
| | | Wohnhäuser | Wohnungen | Zimmer | Küchen | Zimmer | Küchen | | |
| 1 | Akmenes | 1 | 3 | 3 | 3 | 41.4 | 20.60 | 22 | 1.88 |
| 2 | Aldonos | 11 | 22 | 49 | 20 | 464.1 | 158.85 | 160 | 2.92 |
| 3 | Algimanto | 4 | 6 | 10 | 5 | 126.- | 35.30 | 41 | 3.57 |
| 4 | Ariogalos | 51 | 147 | 305 | 98 | 3539.95 | 605.45 | 1100 | 3.22 |
| 5 | Aukuro | 2 | 4 | 4 | – | 65.6 | — | 21 | 3.09 |
| 6 | Bajorų | 12 | 40 | 81 | 33 | 975.8 | 247.- | 335 | 3.95 |
| 7 | Brolių | 4 | 13 | 18 | 10 | 198.75 | 61.40 | 77 | 3.24 |
| 8 | Dailidžių | 9 | 23 | 30 | 18 | 387.50 | 127.50 | 141 | 3.04 |
| 9 | Demokratų | 2 | 11 | 19 | 2 | 249.6 | 52.- | 77 | 3.24 |
| 10 | Dvaro | 23 | 57 | 102 | 41 | 1157.09 | 255.65 | 377 | 3.07 |
| 11 | Erivilkos | 5 | 9 | 16 | 5 | 176.45 | 23.80 | 64 | 2.76 |
| 12 | Gimbuto | 10 | 49 | 83 | 17 | 1068.- | 217.50 | 343 | 3.11 |
| 13 | Girucio | 4 | 14 | 23 | 12 | 248.3 | 65.70 | 85 | 2.92 |
| 14 | Goštauto | 4 | 14 | 30 | 13 | 407.2 | 98.60 | 141 | 2.09 |
| 15 | Griniaus | 31 | 83 | 123 | 65 | 1413 | 542.60 | 533 | 2.65 |
| 16 | Jauniučio | 10 | 26 | 37 | 14 | 448.1 | 102.50 | 151 | 2.96 |
| 17 | Ješiboto | 28 | 68 | 153 | 49 | 1592.15 | 290.- | 610 | 2.61 |
| 18 | Joniškelio | 2 | 3 | 5 | 3 | 49.- | 13.60 | 20 | 2.45 |
| 19 | Koklių | 22 | 67 | 116 | 40 | 1332.24 | 297.06 | 443 | 3.- |
| 20 | Kriščiukaičio | 102 | 232 | 568 | 256 | 6425.30 | 850.35 | 2238 | 2.87 |
| 21 | Kriukų | 14 | 32 | 65 | 17 | 637.80 | 95.20 | 232 | 2.80 |
| 22 | Linkmenų | 9 | 22 | 41 | 11 | 432.80 | 82.40 | 152 | 2.84 |
| 23 | Linkuvos | 56 | 197 | 463 | 124 | 5889.27 | 799.65 | 1840 | 3.20 |
| 24 | Liubarto | 4 | 16 | 18 | 15 | 222.8 | 93.30 | 83 | 2.68 |
| 25 | Liutovaro | 9 | 23 | 29 | 11 | 350.8 | 79.50 | 139 | 2.51 |
| 26 | Margio | 18 | 56 | 81 | 36 | 987.- | 194.50 | 322 | 3.06 |
| 27 | Mėsininkų | 15 | 37 | 86 | 22 | 979.85 | 123.- | 320 | 3.06 |
| 28 | Mildos | 7 | 32 | 67 | 31 | 942.10 | 124.- | 263 | 3.58 |
| 29 | Mindaugo | 6 | 16 | 34 | 15 | 408.85 | 102.- | 127 | 3.22 |
| 30 | Mokyklos | 18 | 35 | 85 | 20 | 846.70 | 126.40 | 298 | 2.84 |
| 31 | Našlaičių | 7 | 17 | 25 | 14 | 313.10 | 80.40 | 110 | 2.84 |
| 32 | Panerių | 38 | 120 | 241 | 98 | 2563.76 | 598.75 | 945 | 2.71 |
| 33 | Puodžių | 19 | 58 | 105 | 52 | 1200.78 | 884.12 | 378 | 3.17 |
| 34 | Rabinų | 10 | 22 | 45 | 20 | 452.30 | 130.40 | 169 | 2.70 |
| 35 | Ramygalos | 8 | 18 | 36 | 18 | 401.40 | 131.75 | 148 | 2.70 |
| 36 | Sinagogos | 2 | 7 | 17 | 6 | 162.25 | 36.90 | 46 | 3.52 |
| 37 | Sintautų | 2 | 2 | 6 | 2 | 54.90 | 17.40 | 20 | 2.74 |
| 38 | Skirgailos | 4 | 17 | 24 | 13 | 352.20 | 93.- | 122 | 2.88 |
| 39 | Skirmanto | 5 | 12 | 22 | 11 | 256.85 | 116.30 | 84 | 3.06 |
| 40 | Stalių | 11 | 27 | 35 | 17 | 406.50 | 136.20 | 156 | 3.60 |
| 41 | Stulginskio | 21 | 61 | 94 | 51 | 1073.20 | 291.30 | 381 | 2.82 |
| 42 | Varguolės | 1 | 1 | 1 | 1 | 12.- | 6.- | 3 | 4.- |
| 43 | Varnių | 23 | 181 | 338 | 113 | 4673.65 | 904.05 | 1532 | 3.05 |
| 44 | Veliuonos | 7 | 12 | 33 | 11 | 335.30 | 62.- | 109 | 3.26 |
| 45 | Vežėjų | 28 | 78 | 167 | 66 | 1725.45 | 418.89 | 598 | 2.91 |
| 46 | Vidurinė | 13 | 30 | 73 | 21 | 758.50 | 123.75 | 261 | 2.90 |
| 47 | Vygrių | 10 | 33 | 72 | 27 | 837.90 | 182.40 | 275 | 3.04 |
| 48 | Vytenio | 26 | 61 | 97 | 49 | 1057.70 | 355.60 | 397 | 2.64 |
| | | 728 | 2106 | 4175 | 1565 | 48701.14 | 10464.60 | 16489 | 2.95 |

Figure 21.5.
A table compiled by the Housing Office for the yearbook "Slobodka Ghetto 1942" showing street by street the number of dwellings, apartments, rooms, kitchens, square meters of space, population, and area per person. On December 31, 1942, the 16,489 inhabitant of the ghetto averaged fewer than three square meters of living space each. (Page 95.)

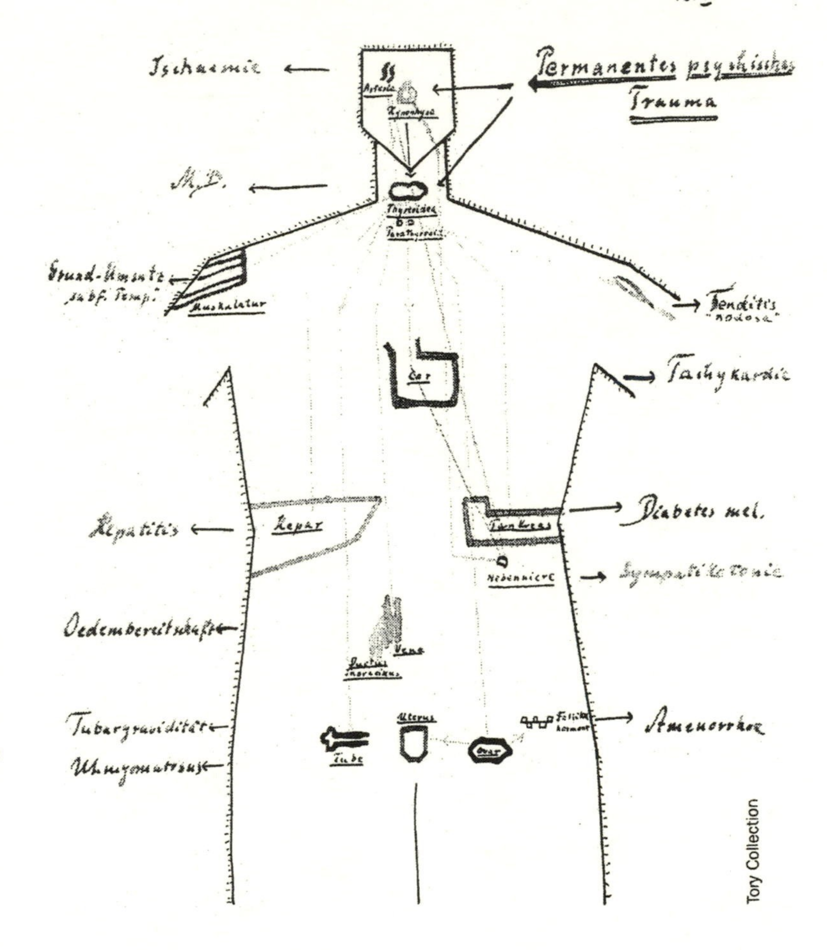


**Figure 21.6.**
Diagram indicating the illnesses that affected women; from a report prepared by Dr. Jacob Nochimowski, chief physician of the ghetto's Labor Office, on "The Effect of Work and of Illnesses on Work in the Ghetto," May 9, 1943. (Page 98.)

in the ghetto that included figure 21.6, depicting the most common areas of concern. Text augmented the graphics:

> During the course of the summer, when clogs were issued, injured feet, prevalent among women and men, took on another form. I shall call this "foot erosion." The tough leather attached to impractical wooden soles with no give and with their hard edges caused deep and painful cuts after a few hours or one day of wear. These wounds took a considerable time to heal.

Figure 21.7.
A first grade class of Jewish children wearing crowns marked with letters from the Hebrew alphabet. This picture was taken at a time prior to the German invasion. Fewer than 5% of Lithuania's Jewish population survived the German occupation. (Page 26.)

Graphics that put specific faces on the numbers have the most evocative effect. Such displays may only be marginally statistical. Combining the photo of a first grade class of Jewish children with the knowledge that only 5% survived creates an effect that a purely statistical display cannot match.

### 21.4. CONCLUSION

Amidst the carnage of World War II, the deliberate extermination of six million Jews stands out in horror and brutality. In order to transform the losses of the Holocaust into something more than a statistic, people have used every medium and genre to convey the stories of individuals, families, and communities: diaries, autobiographies, poems, novels, documentaries, and films.

The Kovno graphs and tables speak volumes about the impact of the Nazi persecution on the ghetto's inhabitants. We see the patterns of life and death as they developed over the months and years and note the categories that were important to the keepers of the records. Disaggregation of some stark totals into meaningful components helps us (however incompletely) not only to comprehend the path to annihilation of a community but also to awaken an emotional response that is somewhat unexpected. It is all the more poignant when we realize that the instrument of our understanding was fashioned, with the express purpose of speaking to future generations, by individuals who fully expected to dissolve into the statistics.

Remembrance without fact is myth; remembrance without emotion is sterile. Worthy memorials draw on both fact and emotion. We should not underestimate the power of even simple numerical displays to help bridge the gap between a statistic and a tragedy.

# PART VI Epilogue

The notion of uncertainty and the role it plays in all human endeavors is very old indeed. "Man plans and God laughs" is an ancient Yiddish expression that exposes clearly how aware the ancients were of the role of uncertainty in everyday lives. This irony parallels Robert Burns' (1759–1796) well-known observation that* "The best-laid plans o' mice an' men gang aft a-gley." The origins of the quantification of uncertainty lie hidden in the depths of time, for surely ancient gamblers must have developed rules of thumb to account for the variation they knew would occur. But the first formal description of the rules of uncertainty can be traced to Jacob Bernoulli (1654–1705), who has been called[1] the father of the quantification of uncertainty. His contributions stand out from the vast array of scientists (e.g., Fermat, Huygens, Leibniz, Arbuthnot) of the time who used the mathematics of permutations to describe the outcomes of games of chance. A twentieth-century take on the Yiddish was Damon Runyon's quantitative observation on uncertainty that "Nothing in life is more than 3 to 1."

The theme of this book is that uncertainty is inevitable and its effects on our lives can be profound. Our chances of navigating this uncertainty successfully are improved, often massively, through (i) understanding the extent of the uncertainty, and (ii) using a few tools to

* From his poem "To a Mouse" in Robert Burns, *Poems, Chiefly in the Scottish Dialect*, Edinburgh: Kilmarnock, 1786.

help us limit the damage that rampant uncertainty can do to our plans. A key tool to understand uncertainty is graphic display, on which I have focused considerable attention. A display that shows the data in all of their variability acts as a control, preventing us from drawing inferences based only on a single summary number (e.g., an average) that can only feebly characterize a complex situation. Using Sam Savage's delicious phrase, graphs can save us from being deceived by the "flaw of averages."

In addition to using clever tools to understand and control uncertainty, to deal with uncertainty successfully we must have a kind of tentative humility. We need a lack of hubris to allow us to see data and let them generate, in combination with what we already know, multiple alternative working hypotheses. These hypotheses are then modified as new data arrive. The sort of humility required was well described by the famous Princeton chemist Hubert N. Alyea, who once told his class, "I say not that it is, but that it seems to be; as it now seems to me to seem to be."

I have tried, with varying degrees of success, to hold my own hubris in check during the telling of the many stories that make up this book. Adopting the mode of story telling was a conscious effort to get the points across. It was chosen in response to the wisdom of the ancient Indian proverb, "Tell me a fact and I'll learn. Tell me a truth and I'll believe. But tell me a story and it will live in my heart forever." Authors can aspire to nothing higher.

Over the entrance to a fifth-century house in North Africa is the inscription,

> Bonus Intra, Melior Exi.*

I could not have said it better.

* Enter good, leave better.

# Notes

## CONTENTS

1. De Moivre, 1730.

## PART I: INTRODUCTION AND OVERVIEW

1. See Wainer, 2005a, chapters 1–3, for more about this remarkable man.

## CHAPTER 1: THE MOST DANGEROUS EQUATION

1. Stigler, 1999, 367–368.
2. This material has previously been reported in Wainer and Zwerling, 2006.
3. For example, Dunn 1977; Beckner, 1983; Howley, 1989; Larson, 1991; Maeroff, 1992; Fowler, 1995.
4. By Schneider, Wysse, and Keesler, and was reported in an article by Debra Viadero in *Education Week* (June 7, 2006).
5. News of the Week in Review, June 18, 2006, p. 2.
6. Kahneman, 2002.

## CHAPTER 2: CURBSTONING IQ AND THE 2000 PRESIDENTIAL ELECTION

1. For example Biemer and Stokes, 1989.
2. Surowiecki, 2004.
3. http://www.museumofhoaxes.com/hoax/weblog/bush_iq/
4. http://nces.ed.gov/nationsreportcard/naepdata

## CHAPTER 3: STUMBLING ON THE PATH TOWARD THE VISUAL COMMUNICATION OF COMPLEXITY

1. See Tufte, 1983/2000; Wainer, 2000b; 2005a.

2. Table no. HS-29, Employment Status of the Civilian Population.

## CHAPTER 5: A POLITICAL STATISTIC

1. http://www.ed.gov/

## CHAPTER 6: A CATCH-22 IN ASSIGNING PRIMARY DELEGATES

1. Previously described by Gelman, Katz, and Tuerlinckx, 2002.

2. See Gelman, 2003, and Gelman, Katz, and Bafumi, 2004.

## CHAPTER 7: TESTING THE DISABLED

1. Bursuck et al., 1989; Yost et al., 1994.

2. Zuriff, 2000, p. 115.

## CHAPTER 8: ETHNIC BIAS OR STATISTICAL ARTIFACT?

1. Eliot, 1987.

2. Herrnstein and Murray, 1994, p. 10.

3. Cleary, 1968.

4. American Educational Research Association, American Psychological Association, National Council on Measurement in Education, 1985, p. 12.

5. For example, Wigdot and Garner, 1982; Vars and Bowen, 1998; Willingham, Pollack, and Lewis, 2000.

6. Cole and Zieky, 2001; Willingham et al., 1990; Willingham and Cole, 1997.

7. Fleming and Garcia, 1998; Pearson, 1993.

8. Leonard and Jiang, 1999; Rebhorn and Miles, 1999.

9. Dimsky, Mittenberg, Quintar, Katell, and Golden, 1998; Elder, 1997.

10. Wainer and Steinberg, 1992.

11. Bok and Bowen, 2002.

12. See, for example, Bowen and Bok, 1998, figure 3.10.

13. Galton, 1889.

14. Wainer, 2000b.

15. Mathews, 2003.

## CHAPTER 9: INSIGNIFICANT IS NOT ZERO

1. CB, 2006, p. 7.

### CHAPTER 11: IMPROVING DATA DISPLAYS

1. Wainer, 1984.

2. Playfair, 1786; 2005.

3. American Society of Mechanical Engineers standards in 1914, 1938, and 1960.

4. Bertin, 1973; Tufte, 1983; 1990; 1996; 2006; Wainer, 2000b; 2005a.

### CHAPTER 12: OLD MOTHER HUBBARD AND THE UNITED NATIONS

1. http://unstats.un.org/unsd/demographic/products/socind/housing.htm

2. Tukey 1977.

### CHAPTER 13: DEPICTING ERROR

1. Presented carefully, with many examples, in Miller, 1966.

2. From Wainer, 1996.

3. For the full story see Wainer, 1996.

4. Variations on this theme were explored earlier by McGill, Tukey, Larsen, 1978, in their notched box plots.

5. Bertin, 1973; Cleveland, 1994; Wainer, 1984, 1992.

6. This idea is merely the visual counterpart to what is commonly done already in many analytic estimation procedures, when we sometimes examine the diagonal elements of the Hessian (the standard errors of the parameters) and sometimes the inverse Hessian (the information matrix).

7. Roberto Bachi's *Graphical Rational Patterns*, 1978.

8. Hoaglin and Tukey, 1985.

9. Fisher, Corbet, and Williams, 1943.

10. For arguments and illustrations of the efficacy and importance of rounding, see Ehrenberg, 1977; Walker and Durost, 1936; Wainer, 1993, 1996.

11. After Herrnstein & Murray, 1994.

12. MacEachren 1992, p. 14; 1994, p. 81.

### CHAPTER 14: THE MENDEL EFFECT

1. Fisher, 1936.

2. van der Waerden, 1968.

3. In Dyson, 2004.

4. Feller, 1968.

### CHAPTER 15: TRUTH IS SLOWER THAN FICTION

1. Galton, 1892; 1893; 1895.

2. See Wainer, 2005a, chapter 8.

3. Stigler, 1999.

### CHAPTER 16: GALTON'S NORMAL

1. Stigler, 1986.

### CHAPTER 17: NOBODY'S PERFECT

1. Friendly and Denis, 2005.
2. Funkhouser and Walker, 1935; Funkhouser, 1937.
3. Friendly and Denis, 2005.
4. Moore, 1911.

### CHAPTER 18: WHEN FORM VIOLATES FUNCTION

1. See Friendly, 2002, Tufte, 1983, or Wainer, 2006, for a detailed story of the plot.
2. Funkhouser, 1937.
3. Tufte, 1983.

### CHAPTER 19: A GRAPHICAL LEGACY OF CHARLES JOSEPH MINARD

1. Available from Graphics Press, Box 430, Cheshire, CT 06410.
2. Tufte, 1983, p. 40.

### CHAPTER 21: NUMBERS AND THE REMEMBRANCE OF THINGS PAST

1. All the included displays are drawn from United States Memorial Council & United State Holocaust Memorial Museum, *Hidden History of the Kovno Ghetto*, Boston: Little, Brown and Company, 1997.

### PART VI: EPILOGUE

1. Stigler, 1986, p. 63.

# References

American Educational Research Association, American Psychological Association, and National Council on Measurement in Education. (1985). *Standards for Educational and Psychological Testing*. Washington, D.C.: American Psychological Association.

———. (1999). *Standards for Educational and Psychological Testing*. Washington, D.C.: American Psychological Association.

Author (2005). *2005 College-Bound Seniors: Total Group Profile Report*. New York: College Board.

Bachi, R. (1978). Proposals for the development of selected graphical methods. In *Graphic Presentation of Statistical Information: Papers Presented at the 136th Annual Meeting of the American Statistical Association*. Washington, D.C.: U.S. Dept. of Commerce, Bureau of the Census.

Beckner, W. (1983). *The Case for the Smaller School*. Bloomington, Ind.: Phi Delta Kappa Educational Foundation.

Benjamini, Y., and Hochberg, Y. (1995). Controlling the False Discovery Rate: A Practical and Powerful Approach to Multiple Testing. *Journal of the Royal Statistical Society, B*, 57, 289–300.

Berry, S. M. (2002). One Modern Man or 15 Tarzans? *Chance*, 15(2), 49–53.

Bertin, J. (1969). *La Semiologie Graphique*. Paris: Mouton.

———. (1973). *Semiologie Graphique*, 2nd Ed. The Hague: Mouton-Gautier. In English, *Semiology of Graphics*, translated by William Berg and Howard Wainer, Madison, Wisconsin: University of Wisconsin Press, 1983.

Bertin, J. (1977). *La Graphique et la Traitement Graphique de l'Information* Paris: Flammarion. In English, *Graphics and Graphic Information Processing,* translated by William Berg; Howard Wainer, Technical Editor. Elmsford, N. Y. : Walter de Gruyter, 1981.

Biemer, P. P., and Stokes, S. L. (1989). The Optimal Design of Quality Control Samples to Detect Interviewer Cheating. *Journal of Official Statistics,* 5(1), 23–39.

Bonferroni, C. E. (1935). Il Calcolo delle Assicurazioni su Gruppi di Teste. In *Studii in Onore del Prof. Salvatore Ortu Carboni.* Rome. Pages 13–60.

Bowen, W. G., and Bok, D. (1998). *The Shape of the River.* Princeton, N.J.: Princeton University Press.

Braun, H., and Wainer, H. (2004). Numbers and the Remembrance of Things Past. *Chance,* 17(1), 44–48.

Bureau of the Census (2004). *Statistical Abstract of the United States 2004–5.* Washington, D.C.: Government Printing Office (http://www.census.gov/prod/www/statistical-abstract-04.html).

Bursuck, W. D., Rose, E., Cowen, S., and Yahaya, M. A. (1989). Nationwide Survey of Postsecondary Education Services for College Students with Learning Disabilities. *Exceptional Children,* 56, 236–245.

Carnevale, A. (1999). Strivers. *Wall Street Journal,* August 31.

Chernoff, H., and Lehman, E. L. (1954). The Use of Maximum Likelihood Estimates in Chi-Square Tests for Goodness of Fit. *Annals of Mathematical Statistics,* 25(3), 579–586.

Chevallier, V. (1871). Notice Nécrologique sur M. Minard, Inspecteur Général des Ponts et Chausées, en Retraite. *Annales des Ponts et Chausées,* 2 (Ser. 5, No. 15), 1–22.

Cleary, A. (1968). Test Bias: Prediction of Grades of Negro and White Students in Integrated Colleges. *Journal of Educational Measurement,* 5, 115–124.

Cleveland, W. S. (1994), *The Elements of Graphing Data,* 2nd Ed. Summit, N.J.: Hobart.

Cleveland, W. S., and McGill, R. (1984). Graphical Perception: Theory, Experimentation, and Application to the Development of Graphical Methods. *Journal of the American Statistical Association,* 79, 531–554.

Cole, N. S., and Zieky, M. J. (2001) The New Faces of Fairness. *Journal of Educational Measurement,* 38(4), 369–382.

Commission on Behavioral and Social Sciences and Education. (2001). *The 2000 Census: Interim Assessment.* Washington, D.C.: National Academies Press.

Daston, L., and Galison, P. (1992). The Image of Objectivity. *Representations,* 40, 81–128.

De Moivre, A. (1730). *Miscellanea Analytica*. London: Tonson and Watts.

Dimsky, Y. I., Mittenberg, W., Quintar, B., Katell, A. D., and Golden, C. J. (1998). Bias in the Use of Standard American Norms with Spanish Translations of the Wechsler Memory Scale—Revised. *Assessment*, 5, 115–121.

Dreifus, C. (2006). A Conversation with Christiane Nüsslein-Volhard. *New York Times*, July 4, F2.

Dunn, F. (1977). Choosing Smallness. In J. Sher (Ed.), *Education in Rural America: A Reassessment of Conventional Wisdom*. Boulder, Colo.: Westview Press.

Dyson, F. (2004). A Meeting with Enrico Fermi, *Nature* 427, 297.

Ehrenberg, A.S.C. (1977). Rudiments of Numeracy. *Journal of the Royal Statistical Society, Series A*, 140, 277–297.

Elder, C. (1997). What Does Test Bias Have to do with Fairness? *Language Testing*, 14, 261–277.

Elliott, R. (1987). *Litigating Intelligence*. Dover, Mass.: Auburn House.

Feller, W. (1968). *An Introduction to Probability Theory and Its Applications*, 3rd Ed. Vol. 1. New York: Wiley.

Fisher, R. A. (1936). Has Mendel's Work Been Rediscovered? *Annals of Science*, 1, 115–137.

Fisher, R. A., Corbet, A. S., and Williams, C. B. (1943). The Relation between the Number of Species and the Number of Individuals in a Random Sample of an Animal Population. *Journal of Animal Ecology*, 12, 42–58.

Fleming, J., and Garcia, N. (1998). Are Standardized Tests Fair to African Americans? Predictive Validity of the SAT in Black and White Institutions. *Journal of Higher Education*, 69, 471–495.

Fowler, W. J., Jr. (1995). School size and student outcomes. Pages 3–25 in H. J. Walberg (Series Ed.) and B. Levin, W. J. Fowler, Jr., and H. J. Walberg (Vol. Eds.), *Advances in Education Productivity*. Vol. 5, *Organizational Influences on Productivity*. Greenwich, Conn.: Jai Press.

Freedle, R. O. (2003). Correcting the SAT's Ethnic and Social-Class Bias: A Method for Reestimating SAT Scores. *Harvard Educational Review*, 73(1), 1–43.

Friendly, M. (2002). Visions and Re-Visions of Charles Joseph Minard. *Journal of Educational and Behavioral Statistics*, 27(1),. 31–51.

Friendly, M., and Denis, D. (2005). The Early Origins and Development of the Scatterplot? *Journal of the History of the Behavioral Sciences*, 41, 103–130.

Friendly, M., and Wainer, H. (2004). Nobody's Perfect. *Chance*, 17(2), 48–51.

Funkhouser, H. G. (1937). Historical Development of the Graphic Representation of Statistical Data. *Osiris*, 3, 269–404.

Funkhouser, H. G., and Walker, H. M. (1935). Playfair and His Charts, *Economic History*, 3, 103–109.

Galton, F. (1869). *Hereditory Genius: An Inquiry into Its Laws and Consequences*. Macmillan: London.

———. (1889). *Natural Inheritance*. London: Macmillan.

———. (1892). *Finger Prints*. London: Macmillan.

———. (1893). *The Decipherment of Blurred Fingerprints*. London: Macmillan.

———. (1895). *Fingerprint Directories*. London: Macmillan.

Geballe, B. (2005). Bill Gates' Guinea Pigs. *Seattle Weekly*, 1–9.

Gelman, A. (2003). Forming Voting Blocs and Coalitions as a Prisoner's Dilemma: A Theoretical Explanation for Political Instability. *Contributions to Economic Analysis and Policy*, 2(1), article 13.

Gelman, A., Katz, J. N., and Bafumi, J. (2004). Standard Voting Power Indexes Don't Work: An Empirical analysis. *British Journal of Political Science*, 34, 657–674.

Gelman, A., Katz, J. N., and Tuerlinckx, F. (2002). The Mathematics and Statistics of Voting Power. *Statistical Science*, 17(4), 420–435.

Gelman, A., and Nolan, D. (2002). *Teaching Statistics: A Bag of Tricks*. Oxford: Oxford University Press.

Goddard, H. H. (1913). The Binet Tests in Relation to Immigration. *Journal of Psycho-Asthenics*, 18, 105–107.

Harness, H. D. (1837). *Atlas to Accompany the Second Report of the Railway Commissioners, Ireland*. Dublin: Irish Railway Commission.

Hauser, R. M., and Featherman, D. L. (1976). Equality of Schooling: Trends and Prospects. *Sociology of Education*, 49, 92–112.

Herrnstein, R. J., and Murray, C. (1994). *The Bell Curve: Intelligence and Class Structure in American Life*, New York: Free Press.

Herschel, J.F.W. (1833). On the Investigation of the Orbits of Revolving Double Stars. *Memoirs of the Royal Astronomical Society*, 5, 171–222.

Hoaglin, D. C., and Tukey, J. W. (1985) Checking the Shape of Discrete Distributions. Chapter 9, pp. 345–416, in David C. Hoaglin, Frederick Mosteller, and John W. Tukey (Eds.), *Exploring Data Tables, Trends and Shapes*. New York: Wiley.

Howley, C. B. (1989). Synthesis of the Effects of School and District Size: What Research Says about Achievement in Small Schools and School Districts. *Journal of Rural and Small Schools*, 4(1), 2–12.

http://www.collegeboard.com/about/news_info/cbsenior/yr2005/links.html

http://www.edwardtufte.com (NEW section) contains a reproduction of Minard's graphic of Hannibal's March as well as an English translation of Chevallier's obituary of Minard, his father-in-law.

Kahneman, D. (2002). Maps of Bounded Rationality: A Perspective on Intuitive Judgment and Choice. Nobel Prize Lecture, December 8, Stockholm, Sweden. http://nobelprize.org/nobel_prizes/economics/laureates/2002/kahneman-lecture.html

Keller, J. B. (1977). A Theory of Competitive Running. Pages 172–178 in S. P. Ladany and R. E. Machol (Eds.), *Optimal Strategies in Sports.* North-Holland: The Hague.

Kierkegaard, S. (1986). *Either/or.* New York: Harper & Row.

Laplace, Pierre Simon (1810). Mémoire sur les Approximations des Formulas qui Sont Functions de Trés-Grand Nombres, et sur Leur Application aux Probabilities. *Mémoires de la Classe des Sciences Mathématiques et Physiques de l'Institut de France, Année 1809,* 353–415, *Supplément,* 559–565.

Larson, R. L. (1991). Small Is Beautiful: Innovation from the Inside Out. *Phi Delta Kappan,* March, 550–554.

Leonard, D. K., and Jiang, J. (1999). Gender Bias and the Predictions of the SATs: A Cry of Despair. *Research in Higher Education,* 40, 375–407.

Lewis, H. (2002). Olympic Performance and Population. *Chance,* 15(4), 3.

Lord, F. M., and Novick, M. R. (1968). *Statistical Theories of Mental Test Scores.* Reading, Mass.: Addison-Wesley.

MacEachren, A. M. (1992). Visualizing Uncertain Information. *Cartographic Perspectives,* 13, 10–19.

———. (1994). Visualization Quality and the Representation of Uncertainty. Chapter 4 in *Some Truth with Maps: A Primer on Symbolization and Design.* Washington, D.C.: Association of American Cartographers.

Maeroff, G. I. (1992). To Improve Schools, Reduce Their Size. *College Board News,* 20(3), 3.

Mann, C. C. (1993) How Many Is too Many? *Atlantic Monthly,* February, http://www.theatlantic.com/issues/93feb/mann1.htm

Marey, E. J. (1878). *La Méthode Graphique dans les Sciences Expérimentales et Particuliérement en Physiologie et en Médecine.* Paris: Masson.

Mathews, J. (2003) The Bias Question. *The Atlantic Monthly,* November, 130–140.

McGill, R., Tukey, J. W., and Larsen, W. (1978). Variations of Box Plots. *The American Statistician,* 32(1), 12–16.

Meserole, M. (Ed.) (1993). *The 1993 Information Please Sports Almanac.* Boston: Houghton Mifflin.

Miller, R. G. (1966). *Simultaneous Statistical Inference.* New York: McGraw-Hill.

Moore, H. L. (1911). *Laws of Wages: An Essay in Statistical Economics,* New York: The Macmillan Co. Reprinted Augustus M. Kelley, New York, 1967.

Mosteller, F., and Tukey, J. W. (1968). Data analysis, Including Statistics. Pages 80–203 in G. Lindzey and E. Aronson (Eds.), *The Handbook of Social Psychology*, Vol. 2. Reading, Mass.: Addison-Wesley.

Patel, M. Dorman, K. S., Zhang, Y-H., Huang, B-L., Arnold, A. P., Sinsheimer, J. S., Vilain, E., and McCabe, E. R. B. (2001). Primate *DAX1, SRY*, and *SOX9*: Evolutionary Stratification of Sex-Determination Pathway. *American Journal of Human Genetics, 68*, 275–280.

Pearson, B. Z. (1993). Predictive Validity of the Scholastic Aptitude Test (SAT) for Hispanic Bilingual Students. *Hispanic Journal of Behavioral Sciences*, 15, 342–356.

Pearson, K. (1892). *The Grammar of Science*. London: Walter Scott.

Pearson, K. P. (1930). *The Life, Letters and Labors of Francis Galton*. Volume III. Cambridge: Cambridge University Press.

Petermann, A. (1852). Distribution of the Population. *Census of Great Britain, 1851*. Lithograph, crayon shading in the British Library.

Pfeffermann, D., and Tiller, R. (2006). Small-Area Estimation with State-Space Models Subject to Benchmark Constraints. *Journal of the American Statistical Association, 101*, 1387–1397.

Playfair, W. (1801). *The Commercial and Political Atlas, Representing, by means of Stained Copper-Plate Charts, The Progress of the Commerce, Revenues, Expenditure, and Debts of England, during the Whole of the Eighteenth Century*. Facsimile reprint edited and annotated by Howard Wainer and Ian Spence. New York: Cambridge University Press, 2006.

———. (1801/2005). *The Statistical Breviary; Shewing on a Principle Entirely New, the Resources of Every State and Kingdom in Europe; Illustrated with Stained Copper-Plate Charts, Representing the Physical Powers of Each Distinct Nation with Ease and Perspicuity*. Edited and introduced by Howard Wainer and Ian Spence. New York: Cambridge University Press.

———. (1821). *A Letter on Our Agricultural Distresses, Their Causes and Remedies*. London: W. Sams.

Quetelet, A. (1832). Recherches sur le Penchant au Crime aux Différens Ages, *Nouveaux Mémoires de l'Académie Royales des Sciences et Belle-Lettres de Bruxelles*, 7, 1–87.

Ramist, L., Lewis, C., and McCamley-Jenkins, L. (1994). *Student Group Differences in Predicting College Grades: Sex, Language and Ethnic Group*. New York: The College Board.

Rebhorn, L. S., and Miles, D. D. (1999). High-Stakes Testing: Barrier to Gifted Girls in Mathematics and Science. *School Science and Mathematics*, 99, 313–319.

Reckase, M. D. (2006). Rejoinder: Evaluating Standard Setting Methods Using Error Models Proposed by Schulz. *Educational Measurement: Issues and Practice*, 25(3), 14–17.

Robinson, A. H. (1982). *Early Thematic Mapping in the History of Cartography*. Chicago: University of Chicago Press.

Rogosa, D. (2005). "A School Accountability Case Study: California API Awards and the *Orange County Register*. Margin of Error Folly." Pages 205–226 in R. P. Phelps (Ed.), *Defending Standardized Testing*. Hillsdale, N. J.: Lawrence Erlbaum Associates.

Rosen, G. (February 18, 2007). Narrowing the Religion Gap. *New York Times Sunday Magazine*, February 18, 11.

Rosenbaum, S., Skinner, R. K., Knight, I. B., and Garrow, J. S. (1985). A Survey of Heights and Weights of Adults in Great Britain. *Annals of Human Biology*, *12(2)*, 115–127.

Savage, S. (in press). *The Flaw of Averages*. New York: John Wiley & Son.

Schmid, C. F. (1983). *Statistical Graphics: Design Principles and Practices*. New York: John Wiley & Son.

Smith, R. L. (1988). Forecasting Records by Maximum Likelihood. *Journal of the American Statistical Association*, 83, 331–338.

Stein, T. (2003). AOL's Death March: The New Media Giant's Internet Offensive Recalls Napoleon's Ill-Fated Russian Campaign. *Wired*, November, 62–63.

Stigler, S. (1986). *The History of Statistics: The Measurement of Uncertainty Before 1900*. Cambridge, Mass.: Harvard University Press.

———. (1987). The Science of Uncertainty. The 404th Convocation Address at The University of Chicago. *The University of Chicago Record*, 15 October, 22(1), 4–5.

———. (1997). Regression Toward the Mean, Historically Considered. *Statistical Methods in Medical Research*, *6*, 103–114.

———. (1999). *Statistics on the Table*. Cambridge, Mass.: Harvard University Press.

Surowiecki, J. (2004). *The Wisdom of Crowds*. New York: Doubleday.

Thoreau, H. D. (1968). *The Writings of Henry David Thoreau: Journal, Vol. 2, 1850–September 15, 1851*. Bradford Torrey (Ed.). New York: AMS Press. From the 1906 edition.

Tryfos, P., and Blackmore, R. (1985). Forecasting Records. *Journal of the American Statistical Association.*, 80, 46–50.

Tufte, E. R. (1983/2000). *The Visual Display of Quantitative Information*. Cheshire, Conn.: Graphics Press.

Tufte, E. R. (1990). *Envisioning Information*. Cheshire, Conn.: Graphics Press.

———. (1996). *Visual Explanations*. Cheshire, Conn.: Graphics Press.

———. (2006). *Beautiful Evidence*. Cheshire, Conn.: Graphics Press.

Tukey, J. W. (1977). *Exploratory Data Analysis*. Reading, Mass.: Addison-Wesley.

United States Memorial Council & United State Holocaust Memorial Museum (1997). *Hidden History of the Kovno Ghetto*. Boston: Little, Brown and Company. It is from this source that all the included displays are drawn.

Unwin, A., Theus, M., and Hofmann, H. (2006). *Graphics of Large Datasets: Visualizing a Million*.Springer-Verlag: New York.

U.S. Bureau of the Census. (1986). Memorandum for General Distribution from I. Schreiner and K. Guerra, Subject "Falsification Study" (September 1982 through August 1985).

van der Waerden, B. L. (1968). Mendel's Experiments. *Centaurus*, 12, 275–288.

Van Dorn, W. G. (1991). Equations for Predicting Record Human Performance in Cycling, Running and Swimming *Cycling Science*, September & December, 13–16.

Vars, F., and Bowen, W. (1998). Scholastic Aptitude Test Scores, Race, and Academic Performance in Selective Colleges and Universities. In C. Jencks and M. Phillips (Eds.), *The Black-White Test-Score Gap*. Washington, D.C.: Brookings Institution.

Vasilescu, D., and Wainer, H. (2005). Old Mother Hubbard and the United Nations: An Adventure in Exploratory Data Analysis (with Discussion). *Chance*, 18(3), 38–45.

Viadero, D. (2006). Smaller not Necessarily Better, School-Size Study Concludes. *Education Week*, June 7, 25(39), 12–13.

Wainer, H. (1984). How to Display Data Badly. *American Statistician*, 38, 137–147.

———. (1992). Understanding Graphs and Tables. *Educational Researcher*, 21, 14–23.

———. (1993). Tabular Presentation. *Chance*, 6(3), 52–56.

———. (1996). Depicting Error. *American Statistician*, 50(2), 101–111.

———. (1997). Improving Tabular Displays: With NAEP Tables as Examples and Inspirations. *Journal of Educational and Behavioral Statistics*, 22, 1–30.

———. (2000a). Kelley's Paradox. *Chance*, 13(1), 47–48.

———. (2000b). *Visual Revelations: Graphical Tales of Fate and Deception from Napoleon Bonaparte to Ross Perot*. 2nd Ed. Hillsdale, N. J.: Lawrence Erlbaum Associates.

———. (2001). New Tools for Exploratory Data Analysis: III. Smoothing and Nearness Engines. *Chance*, 14(1), 43–46.

———. (2003c). La Diffusion de Quelques Idées: A Master's Voice. *Chance*, 16(3), 58–61.

———. (2003a). A Graphical Legacy of Charles Joseph Minard: Two Jewels from the Past. *Chance*, 16(1), 56–60.

———. (2003b). How Long is Short? *Chance*, 16(2), 55–57.

———. (2004). Curbstoning IQ and the 2000 Presidential Election. *Chance*, 17(4), 43–46.

———. (2005a). *Graphic Discovery: A Trout in the Milk and Other Visual Adventures*. Princeton, N.J.: Princeton University Press.

———. (2005b). Stumbling on the Path Toward the Visual Communication of Complexity. *Chance*, 18(2), 53–54.

———. (2006). Using Graphs to make the Complex Simple: The Medicare Drug Plan as an Example, *Chance*, 19(2), 55–56.

———. (2007). Galton's Normal Is too Platykurtic. *Chance*, 20(2), 57–58.

———. (2007). Improving Data Displays: The Media's and Ours. *Chance*, 20(3), 8–16.

———. (2007). Insignificant Is not Zero: Rescoring the SAT as an Example. *Chance*, 20(1), 55–58.

———. (2007). The Most Dangerous Equation. *American Scientist*, 95 (May–June), 249–256.

Wainer, H., and Brown, L. (2007). Three Statistical Paradoxes in the Interpretation of Group Differences: Illustrated with Medical School Admission and Licensing . Chapter 26, pp. 893–918, in C. R. Rao and S. Sinharay (Eds.), *Handbook of Statistics*, Vol. 27, *Psychometrics*. Elsevier Science, Amsterdam.

Wainer, H., Bridgeman, B., Najarian, M., and Trapani, C. (2004). How Much Does Extra Time on the SAT Help? *Chance*, 17(2), 19–24.

Wainer, H., and Clauser, B. (2005). Truth is Slower than fiction: Francis Galton as an illustration. *Chance*, 18(4), 52–54.

Wainer, H., and Friendly, M. (2004). Nobody's Perfect. *Chance*, 17(2), 48–51.

Wainer, H., and Gelman, A. (2007). A Catch-22 in Assigning Primary Delegates. *Chance*, 20(4), 6–7.

Wainer, H. , Gesseroli, M., and Verdi, M. (2006). Finding What Is not There Through the Unfortunate Binning of Results: The Mendel Effect, *Chance*, 19(1), 49–52.

Wainer, H., and Koretz, D. (2003). A Political Statistic. *Chance*, 16(4), 45–47.

Wainer, H., Njue, C., and Palmer, S. (2000). Assessing Time Trends in Sex Differences in Swimming and Running (with Discussions). *Chance*, 13(1), 10–15.

Wainer, H., and Skorupski , W. P. (2005). Was It Ethnic and Social-Class Bias or Was It Statistical Artifact? Logical and Empirical Evidence against Freedle's Method for Reestimating SAT scores. *Chance*, 18(2), 17–24.

Wainer, H., and Steinberg, L. S. (1992). Sex Differences in Performance on the Mathematics Section of the Scholastic Aptitude Test: A Bidirectional Validity Study. *Harvard Educational Review*, 62, 323–336.

Wainer, H., and Zwerling, H. (2006). Logical and Empirical Evidence that Smaller Schools do not Improve Student Achievement. *Phi Delta Kappan*, 87, 300–303.

Walker, H. M., and Durost, W. N. (1936). *Statistical Tables: Their Structure and Use*. New York: Bureau of Publications, Teachers College, Columbia University.

Wigdor, A., and Garner, W. (Eds.). (1982). *Ability Testing: Uses, Consequences, and Controversies*. Washington, D.C.: National Academy Press.

Williams, V.S.L., Jones, L. V., and Tukey, J. W. (1994). *Controlling Error in Multiple Comparisons with Special Attention to the National Assessment of Educational Progress*. Technical Report No. 33. Research Triangle Park, N.C.: National Institute of Statistical Sciences.

Willingham, W. W., and Cole, N. S. (1997). *Gender and Fair Assessment*. Mahwah, N.J.: Lawrence Erlbaum Associates.

Willingham, W. W., Lewis, C. , Morgan, R., and Ramist, L. (1990). *Predicting College Grades: An Analysis of Institutional Trends over Two Decades*. New York: The College Board.

Willingham, W. W., Pollack, J. M., and Lewis, C. (2000). *Grades and Test Scores: Accounting for Observed Differences*. Princeton, N.J.: Educational Testing Service.

Yost, D. S., Shaw, S. F., Cullen, J. P., and Bigaj, S. J. (1994). Practices and Attitudes of Postsecondary LD Service Providers in North America. *Journal of Learning Disabilities*, 27, 631–640.

Zuriff, G. E. (2000). Extra Examination Time for Students with Learning Disabilities: An Examination of the Maximum Potential Thesis. *Applied Measurement in Education*, 13, 99–117.

# Source Material


**CHAPTER 1**

Wainer, H. (2007). The Most Dangerous Equation. *American Scientist*, 95 (May–June), 249–256.

**CHAPTER 2**

Wainer, H. (2004). Curbstoning IQ and the 2000 Presidential Election. *Chance*, 17(4), 43–46.

**CHAPTER 3**

Wainer, H. (2005). Stumbling on the Path Toward the Visual Communication of Complexity. *Chance*, 18(2), 53–54.

**CHAPTER 4**

Wainer, H. (2006). Using Graphs to Make the Complex Simple: The Medicare Drug Plan as an Example. *Chance*, 19(2), 55–56.

**CHAPTER 5**

Wainer, H., and Koretz , D. (2003). A Political Statistic. *Chance*, 16(4), 45–47.

**CHAPTER 6**

Wainer, H., and Gelman, A. (2007). A Catch-22 in Assigning Primary Delegates. *Chance*, 20(4), 6–7.

**CHAPTER 7**

Wainer, H. (2000). Testing the Disabled: Using Statistics to Navigate between the Scylla of Standards and the Charybdis of Court Decisions. *Chance*, 13(2), 42–44.

**CHAPTER 8**

Wainer, H., and Skorupski , W. P. (2005). Was it Ethnic and Social-Class Bias or Was it Statistical Artifact? Logical and Empirical Evidence against Freedle's Method for Reestimating SAT Scores . *Chance*, 18(2), 17–24.

**CHAPTER 9**

Wainer, H. (2007). Insignificant Is not Zero: Rescoring the SAT as an Example. *Chance*, 20(1), 55–58.

**CHAPTER 10**

Wainer, H. (2003). How Long Is Short? *Chance*, 16(2), 55–57.

**CHAPTER 11**

Wainer, H. (2007). Improving Data Displays: The Media's and Ours. *Chance*, 20(3), 8–16.

**CHAPTER 12**

Vasilescu, D., and Wainer, H. (2005). Old Mother Hubbard and the United Nations: An Adventure in Exploratory Data Analysis (with Discussion). *Chance*, 18(3), 38–45.

**CHAPTER 13**

Wainer, H. (1996). Depicting Error. *American Statistician*, 50(2), 101–111.

**CHAPTER 14**

Wainer, H. , Gesseroli, M., and Verdi, M. (2006). Finding What Is not there Through the Unfortunate Binning of Results: The Mendel Effect, *Chance*, 19(1), 49–52.

**CHAPTER 15**

Wainer, H., and Clauser, B. (2005). Truth is Slower than Fiction: Francis Galton as an Illustration. *Chance*, 18(4), 52–54.

**CHAPTER 16**

Wainer, H. (2007). Galton's Normal Is too Platykurtic. *Chance*, 20(2), 57–58.

**CHAPTER 17**

Wainer, H., and Friendly, M. (2004). Nobody's Perfect. *Chance*, 17(2), 48–51.

**CHAPTER 18**

Wainer, H. (2004). When Form Violates Function. *Chance*, 17(3), 49–52.

**CHAPTER 19**

Wainer, H. (2003). A Graphical Legacy of Charles Joseph Minard: Two Jewels from the Past. *Chance*, 16(1), 56–60.

**CHAPTER 20**

Wainer, H. (2003). La Diffusion de Quelques Idées: A master's Voice. *Chance*, 16(3), 58–61.

**CHAPTER 21**

Braun, H., and Wainer, H. (2004). Numbers and the Remembrance of Things Past. *Chance*, 17(1), 44–48.

# Index